Red Giants and White Dwarfs

In an unbroken sequence of events the universe expands and cools, earth are formed, and life arises on the scene.

extending over 20 billion years, stars are born and die, the sun and on the earth. Finally man appears

Red Giants and

WARNER BOOKS

A Warner Communications Company

White Dwarfs

revised edition

Robert Jastrow

Warner Books Edition

Revised edition copyright © 1979 by Reader's Library, Inc.

Earlier editions copyright © 1967, 1971 by Robert Jastrow

Reprinted by arrangement with the Author
and Reader's Library, Inc.

Cover photo: Barrett Gallagher

W A Warner Communications Company

Warner Books, Inc., 75 Rockefeller Plaza,
New York, N. Y. 10019

Printed in the United States of America

First printing: October 1980

10 9 8 7 6 5 4 3 2 1

Library of Congress Cataloging in Publication Data

Jastrow, Robert
  Red giants and white dwarfs.

  Reprint of the New ed. published by Norton, New
York.
  Includes index.
    1. Astronomy—Popular works.   2.   Life—Origin.
I. Title.
[QB44.J379 1980]   523.1'2   80-14548
ISBN 0-446-97349-1 (U.S.)
ISBN 0-446-97764-0 (Canada)

*To my parents
with love*

Contents

FROM THE PREFACE
TO THE FIRST EDITION

This book had its genesis in a series of fifty-seven half-hour television programs on space science that I did for the CBS network in 1964. I decided to explain in these programs all fields of modern science, because the experiments carried on in the space program involve almost every important scientific problem, and I wanted the audience to understand why these experiments are worthwhile. I spanned most of the scientific spectrum—physics, astronomy, earth science and biology.

Toward the end of the series of programs I had a revelation that had never come to me before in fifteen years as a scientist. I realized that the latest advances in the separate fields of science, fascinating in themselves, are the brightly colored fragments of a mosaic which, when viewed from a distance, forms an image of the human observer himself, and of his origins. Science has uncovered evidence suggesting that we owe our existence to events which took place billions of years ago in stars that lived and died long before our solar system was formed. The scientific story of creation touches on the central problems of man's existence: What am I? How did I get here? What is my relation to the rest of the universe? The ideas are simple and beautiful; they can be expressed in clear language, without the use of jargon or mathematics. The story of man's origins goes far beyond the concepts of Darwin; it begins earlier than the time of our tree-dwelling ancestors, and much earlier than the period, several billion years ago, when the lowest forms of life first appeared on the face of the earth; it crosses the threshold between the living and the non-living worlds and goes back in time to the parent cloud of hydrogen out of which all existing things are descended.

Many colleagues helped me in preparation of this volume. I am particularly indebted to several people who were kind enough to take the time required for a careful reading and the preparation of detailed criticisms of individual chapters. These include Professors Bengt Strömgren and A. G. W. Cameron and Dr. Richard Stothers for the chapters on the birth and death of stars and on cosmology; Professors Harold Urey, Gordon J. F. MacDonald and Paul W. Gast for the chapters on the origin of the solar system, the moon and the history of the earth; Dr. Gordon M. Tomkins and Professors Joshua Lederberg and Stanley L. Miller for the chapter on the origin of life; and Professors Edwin Colbert, John Imbrie, Armand V. Oppenheimer, Colin Pittendrigh and G. G. Simpson for the chapters on evolution.

PREFACE TO THE NEW EDITION

The last ten years have witnessed a remarkable explosion of knowledge in the fields of science covered by the first edition of this work. The new findings relate to some of the most interesting topics in science today: black holes in space, life on Mars, the moons of Jupiter and Saturn and new evidence for the creation of the universe.

The entire book has been revised and updated to incorporate this recently acquired information. The chapters on the moon and planets, and on UFOs, are largely or entirely new. The discussion of black holes in the chapter on the life story of the stars has been considerably expanded. The descriptions of Mars and Jupiter, and the prospects for life on those planets, are based on the latest information available in 1979. The chapter on Jupiter was adapted in part from my article in the *New York Times Magazine.*

Many new photographs, including a color section, have been added throughout the book. Some, obtained by cameras on the Viking and Voyager spacecraft, reveal extraordinary sights never witnessed by the human eye prior to the last few years.

This book is a companion work to *Until the Sun Dies.* It deals mainly with events in the cosmos leading to the threshold of human existence, while the latter work explores the forces that have shaped human beings into their present form. The two books together are a scientific version of the story of Genesis that unites the new knowledge of the twentieth century with the great ideas of Darwin.

I am indebted to Elizabeth Stevens for valuable assistance in all aspects of editing and production. My mother, Marie Jastrow, made many suggestions which improved the clarity of the book for the general reader, and helped me to escape from the limitations of a professional scientist's writing style. Her contributions appear throughout this edition.

September, 1979

Red Giants and White Dwarfs

1 PROLOGUE

At the age of thirty I began to grow increasingly restless in my work. I had taught theoretical physics at Yale and Columbia, worked in the nuclear physics laboratory of the University of California in Berkeley, and spent some time at the Institute for Advanced Study in Princeton. Now I found myself still engaged in theoretical problems of nuclear structure, this time at the Naval Research Laboratory in Washington. My research had been limited to the fields of atomic and nuclear physics, and I had no feeling for the aspects of science that involved events outside the laboratory. Nonetheless I was intrigued by the activities of a group of scientists who were engaged in a completely different effort in another part of the Naval Research Laboratory. These were people working on Project Vanguard, the first American program for launching an artificial satellite.

The Vanguard project captured my imagination, and I decided to become involved in it if at all possible. With this end in mind I walked over one day to offer my services to J. W. Siry, the chief theoretician of the project. Dr. Siry told me of a problem which he considered to be important, but which he could not take care of himself because of the burdens placed on him by the preparations for the first Vanguard launching. The problem he proposed was to calculate the slowing down of a satellite in orbit as a result of its collisions with the molecules of air. I did the calculations with a collaborator, Dr. C. A. Pearse. Our results suggested some peculiar electrical effects on the motion of the satellite. This research introduced me to the physics of satellite orbits.

Soon the Russians, and not the Americans, launched the first man-made satellite into space. Emboldened by the extraordinary event, I entered the Vanguard control room for the first time, to find the Vanguard radio networks tracking the Sputnik as it circled the earth.

Accurate computing methods for the tracking of satellites were not yet ready, because the American program was geared to the launching of the Vanguard satellite scheduled to go up several months later.

The Vanguard team was under pressure to provide information on the position of the Sputnik. I asked whether I could help out with the preliminary calculations of the orbit. A naval ensign and a mathematics major recently out of Middlebury College were working at desks in a large room. Teletype machines clacked away with the results of radio positions of the Sputnik gathered from United States tracking stations stretched across North and South America. In the course of very rough calculations, I found the expected result that the Sputnik was steadily losing energy as a result of atmospheric friction. From the rate of energy loss the density of the earth's upper atmosphere could be calculated. Through this circumstance I first became acquainted with atmospheric physics.

The work on satellite orbits led to an invitation to participate in an International Rocket and Satellite Conference scheduled to meet in Moscow. At the same time I was given another assignment. Reports had been received of a bright object flying across the sky above Alaska and the western part of the United States, at around the time that the friction of the atmosphere would have brought back to earth the rocket which boosted the Sputnik into its orbit. The Russians announced that the rocket had come down over American territory, and that they wanted it back. Premier Khrushchev personally asked for it. Doubtless the United States government would have been delighted to accede to this request, but they did not have the rocket and were unable to find it.

I examined the available reports and radar sightings on the final orbits of the rocket and came to the conclusion that it had lasted in its orbit for six to eight hours after its final passage over the United States, and that most probably it had come down somewhere along a 2000-mile arc stretching between the eastern part of the Soviet Union and China.

On reaching the conference in Moscow in July, I presented my conclusions on the fate of the rocket. A pin could have been heard to drop in the hall as I delivered the dry summary of my calculations. No questions were asked by the several hundred Russian scientists who were present, and when the foreign correspondents filed their stories the Soviet censor held them up, although they were released the next day. To this day I do not know whether the Soviet scientists found the analysis convincing, or simply assumed that I had been ordered by the American government to manufacture my results.

From the Moscow conference I went on to the second United Nations Conference on the Peaceful Uses of Atomic Energy to discharge what turned out to be my last duty as a nuclear physicist. On returning to the United States in the fall, I was asked to join NASA, the newly created National Aeronautics and Space Administration, which had just been set up by the government in response to the Soviet challenge. My duty was to bring into existence a Theoretical Division of NASA, which was to devote itself to basic research in astronomy and the planetary sciences.

In a general way the area of research of the new division was obvious. It would include all scientific problems that could be studied by instruments carried on satellites and interplanetary rockets. But this definition of the job could have included most of science. The problem was to choose a few important problems and concentrate on them. What were the important problems?

I took my oath of appointment and three weeks later I traveled across the United States to the laboratory at La Jolla, California, to visit a man who, I had been told, would be able to give me some advice. Professor Harold Clayton Urey had written a book on the moon and the planets, and was well known for the intensity of his interest in the scientific study of these objects. On arriving I introduced myself to him and asked him whether he could suggest some problems for the Theoretical Division of NASA to tackle, which would give us experience in a field so new to us.

Professor Urey seemed pleased to be sought out by a physicist working for the new space agency. He sat me down, handed me his book on the planets, opened to the chapter on the moon, and began to tell me of the unique importance which this arid and lifeless body has for anyone who wishes to understand the origin of the earth and other planets. I was fascinated by his story, which had never been told to me before in fourteen years of study and research in physics. Harold Urey has the marvelous quality of an intense, almost childlike curiosity regarding all aspects of the natural world. This kind of curiosity is a rare quality. Through the eyes of Harold Urey, and others whom I met later, I first became acquainted with questions in science that were new to me, and completely different from the laboratory physics that had dominated my earlier training. As a graduate student, my life had revolved around the atom and the nucleus. In my world the laws of physics applied, and all events were governed by the action of the basic forces of nature—gravity, electromagnetism and the nuclear forces. Astronomy and the earth sciences—the study of the impact of these forces on a grand scale—were a closed book to me. Biology was an entirely alien subject.

Now I learned for the first time how stars and planets are born, how the solar system was formed, what conditions may have prevailed on the primitive earth, and how life may have developed on our planet. My previous work in physics had never led me to consider these matters. Yet the fact is that a single thread of evidence runs from the atom and the nucleus through the formation of stars and planets to the complexities of the living organism. Discoveries of the last decades link the world of the nucleus to the world of life in a chain of cause and effect which extends over billions of years, commencing with the formation of stars in our galaxy and ending with the appearance of man on the earth. At the beginning of this history there existed only atoms of the primeval element, hydrogen, which swirled through outer space in vast clouds. These clouds were the raw stuff out of which stars, planets and men were made. Occasionally the atoms of a cloud were drawn together by the attractive forces of gravity; with the passage of time the cloud contracted to a small, dense globe of gas; heated by self-compression, it rose in temperature until, at a level of some millions of degrees, its center

Harold Clayton Urey

burst into nuclear flame. Out of such events, stars were born.

Within the newborn star a series of nuclear reactions set in, in which all the other elements of the universe were manufactured out of the basic ingredient, hydrogen. Eventually these nuclear reactions died out, and the star's life came to an end. Deprived of its resources of nuclear energy, it collapsed under its own weight, and in the aftermath of the collapse an explosion occurred, spraying out to space all the materials that had been created within the star during its lifetime.

In the course of time, new stars, some with planets around them, condensed out of these materials. The sun and the earth were formed in this way, four and a half billion years ago, out of materials manufactured in the bodies of other stars earlier in the life of the Galaxy, and then dispersed to space when those stars exploded.

When the earth was first formed it must have been barren, but within one billion years or so life appeared on its surface. How can we explain this fact? What conditions prevailed on the earth during those first billion years? Recent discoveries hold clues to the answer. The biologists have discovered that certain molecules are the building blocks of all living creatures; they have created these molecules in the laboratory, out of the kinds of chemicals that existed in the atmosphere and oceans of the primitive earth; and they assert that life probably developed on the earth out of such molecules, several billion years ago.

Fossils preserved in the rocks of the earth's crust show that, during the next three billion years, a million varieties of plants and animals evolved out of these first living organisms. Why did those particular plants and animals arise, and not others? What forces pressed life into the forms it now possesses? Darwin's theory of evolution supplies an answer: Throughout the history of the earth the pressures of the struggle for existence have worked steadily on all creatures, shaping and molding the forms of life until each species acquires the best possible chance for survival in its environment. The fossil record displays the results of this process; it shows the gradual proliferation of many varieties of life, each specially adapted to one set of conditions. With the passage of time, and successive changes of climate, new forms of life spring up, and old ones are extinguished. At the end of the long chain of development man appears, as the product of a line of evolution that goes back far beyond his tree-dwelling ancestors, and even beyond the first forms of life on the earth. Man's history began billions of years before the solar system itself was formed; it began in a swirling cloud of primordial hydrogen. This is the history I shall relate in this book.

2 THE SIZE OF THINGS

I once had occasion to testify before the United States Senate Space and Aeronautics Committee on the scientific background of the space program. My talk dealt with the manner in which all substances in the universe are assembled out of neutrons, protons and electrons as the basic building blocks. After I left the chamber a senior NASA official continued with a summary of the major space science achievements of the last year. Apparently my scholarly presentation had perplexed the senators, although they were anxious to understand the concepts I had presented. However, the NASA official's relaxed manner reassured them, and someone asked him: "How big is the electron? How much smaller is it than a speck of dust?" The NASA official correctly replied that the size of an electron is to a dust speck as the dust speck is to the entire earth.

The electron is indeed a tiny object. Its diameter is one 10-trillionth of an inch, a million times smaller than can be seen with the best electron microscope. Its weight is correspondingly small; 10,000 trillion trillion electrons make up one ounce. How can we be certain that such a small object exists? No one has ever picked up an electron with a pair of forceps and said, "Here is one." The evidence for its existence is all indirect. During the 150 years from the late eighteenth century to the beginning of the twentieth century a great variety of experiments were carried out on the flow of electricity through liquids and gases. The existence of the electron was not proved conclusively by any single one of these experiments. However, the majority of them could be explained most easily if the physicist assumed that the electricity was carried by a stream of small particles, each bearing its own electrical charge. Gradually physicists acquired a feeling, bordering on conviction, that the electron actually exists.

The question now was, how large is the electron, and how much electric charge does each electron carry? The clearest answer to this question came from an American physicist, Robert Millikan, who worked on the problem at the University of Chicago in the first decades of the twentieth century. Millikan arranged a device, clever for its simplicity, in which an atomizer created a mist of very fine droplets of oil just above a small hole in the top of a container. A small number of the droplets fell through the hole and slowly drifted to the bottom of the container. Millikan could see the motions of these droplets very clearly by illuminating them from the side with a strong light so that they appeared as bright spots against a dark background. Millikan discovered that some of these droplets carried a few extra electrons, which had been picked up in the atomizing process. By applying an electrical force to the droplets and studying their motions in response to this force, he could deduce the amount of electric charge carried by the electrons on each droplet. This charge turned out to be exceedingly minute. As a demonstration of its minuteness, it takes an electric current equivalent to a flow of one million trillion electrons every second to light a 10-watt bulb. All this happened rather recently in the history of science. Millikan's first accurate measurements were completed in 1914.

The tiny _electron_, and two sister particles, are the building blocks out of which all matter in the world is constructed. The sister particles to the electron are the _proton_ and the _neutron_. They were discovered even more recently than the electron; the proton was identified in 1920 and the neutron was first discovered in 1932. These two particles are massive in comparison with the electron—1840 times as heavy—but still inconceivably light by ordinary standards. The three particles combine in an amazingly simple way to form the objects we see and feel. A strong force of attraction binds neutrons and protons together to form a dense, compact body called the nucleus, whose size is somewhat less than one-trillionth of an inch. Electrons are attracted to the nucleus and circle around it as the planets circle around the sun, forming a solar system in miniature.

5. _Atom_
consists of
electrons &
the nucleus

5

23

1. Neutrons
2. Protons
3. Electrons circle nucleus
4. Neutrons & Protons form the NUCLEUS

Together the electrons and the nucleus make up the *atom.*

The size of a typical atom is one hundred-millionth of an inch. To get a feeling for the smallness of the atom compared to a macroscopic object, imagine that you can see the individual atoms in a kitchen table, and that each atom is the size of a grain of sand. On this scale of enlargement the table will be 2000 miles long.

The comparison of the atom with a grain of sand implies that the atom is a solid object. Actually, the atom consists largely of empty space. Each of the atoms that makes up the surface of a table consists of a number of electrons orbiting around a nucleus. The electrons form a diffuse shell around the nucleus, marking the outer boundary of the atom. The size of the atom is 10,000 times as great as the size of the nucleus at the center. If the outer shell of electrons in the atom were the size of the Astrodome that covers the Houston baseball stadium, the nucleus would be a ping-pong ball in the center of the stadium. That is the emptiness of the atom.

If most of the atom is empty space, why does a tabletop offer resistance when you push it with your finger? The reason is that the surface of the table consists of a wall of electrons, the electrons belonging to the outermost layer of atoms in the tabletop; the surface of your finger also consists of a wall of electrons; where they meet, strong forces of electrical repulsion prevent the electrons in your fingertip from pushing past the outermost electrons in the top of the table into the empty space within each atom. An atomic projectile such as a proton, accelerated to high speed in a cyclotron, could easily pass through these electrons, which are, after all, rather light and unable to hurl back a fast-moving object. But it would take more force than the pressure of the finger can produce to force them aside and penetrate the inner space of the atom.

The concept of the empty atom is a recent development. Isaac Newton described atoms as "solid, massy, hard, impenetrable, moveable particles." Through the nineteenth century, physicists continued to regard them as small, solid objects. Lord Rutherford, the greatest experimental physicist of his time, once said, "I was brought up to look at the atom as a nice hard fellow, red or grey in color, ac-

A photograph of Ernest Rutherford (1871-1937) taken
outside the Cavendish Laboratory in 1934

cording to taste." At the beginning of the twentieth century, J. J. Thomson, a British physicist and one of the pioneers in the investigation of the structure of matter, believed that the atom was a spherical plum pudding of positive electric charge in which negatively charged electrons were embedded like raisins. No one knew that the mass of the atom, and its positive charge, were concentrated in a small, dense nucleus at the center, and that the electrons circled around this nucleus at a considerable distance. But in 1911 Rutherford, acting on a hunch, instructed his assistant, Hans Geiger, and a graduate student named Marsden, to fire a beam of alpha particles into a bit of thin gold foil. These alpha particles are extremely fast-moving atomic projectiles which should have penetrated the gold foil and emerged from the other side. Most of them did, but Geiger and Marsden found that in a very few cases the alpha particles came out of the foil on the same side they had entered. Rutherford said later, "It was quite the most incredible event that has ever happened to me in my life. It was almost as incredible as if you fired a 15-inch shell at a piece of tissue paper and it came back and hit you." Later Geiger told the story of the occasion on which Rutherford saw the meaning of the experiment. He relates: "One day [in 1911] Rutherford, obviously in the best of spirits, came into my room and told me that he now knew what the atom looked like and how to explain the large deflections of the alpha particles." What had occurred, Rutherford had decided, was that now and then an alpha particle hit a massive object in the foil, which bounced it straight back. He realized that the massive objects must be very small since the alpha particles hit them so rarely. He concluded that most of the mass of the atom is concentrated in a compact body at its center, which he named the nucleus. Rutherford's discovery opened the door to the nuclear era.

Let us continue with the description of the manner in which the universe is assembled out of its basic particles. Atoms are joined together in groups to form molecules, such as water, which consists of two atoms of hydrogen joined to one atom of oxygen. Large numbers of atoms or molecules cemented together form solid matter. There are a trillion trillion atoms in a cubic inch of an ordinary solid

H — hydrogen
O — Oxygen

molecule-
of water
formed by
2H Atoms
& 1 O Atom

Atoms

substance, which is roughly the same as the number of grains of sand in all the oceans of the earth.

The earth itself is an especially large collection of atoms bound together in a ball of rock and iron 8000 miles in diameter, weighing six billion trillion tons. It is one of nine planets, which are bound to the sun by the force of gravity. Together the sun and planets form the solar system. The largest of the planets is Jupiter, whose diameter is 86,000 miles; Mercury, the smallest is 3100 miles across, one-third the size of the earth, and scarcely larger than the moon. All the planets are dwarfed by the sun, whose diameter is one million miles. The weight of the sun is 700 times greater than the combined weight of the nine planets. Like the atom, the solar system consists of a massive central body—the sun—surrounded by small, light objects—the planets—which revolve about it at great distances.

The sun is only one among 200 billion stars that are bound together by gravity into a large cluster of stars called the Galaxy. The stars of the Galaxy revolve about its center as the planets revolve about the sun. The sun itself participates in this rotating motion, completing one circuit around the Galaxy in 250 million years.

The Galaxy is flattened by its rotating motion into the shape of a disk, whose thickness is roughly one-fiftieth of its diameter. Most of the stars in the Galaxy are in this disk, although some are located outside it. A relatively small, spherical cluster of stars, called the nucleus of the Galaxy, bulges out of the disk at the center. The entire structure resembles a double sombrero with the galactic nucleus as the crown and the disk as the brim. The sun is located in the brim of the sombrero about three-fifths of the way out from the center to the edge. When we look into the sky in the direction of the disk we see so many stars that they are not visible as separate points of light, but blend together into a luminous band stretching across the sky. This band is called the Milky Way.

The stars within the Galaxy are separated from one another by an average distance of about 36 trillion miles. In order to avoid the frequent repetition of such awkwardly large numbers, astronomical distances are usually expressed in units of the light year. A light year is defined as the distance covered in one year by a ray of light, which travels at 186,000 miles per second. The distance turns out to be six trillion miles; hence in these units the average distance between stars in the Galaxy is five light years, and the diameter of the Galaxy is 100,000 light years.

In spite of the enormous size of our galaxy, its boundaries do not mark the edge of the observable universe. The 200-inch telescope on Mount Palomar has within its range no less than 10 billion other galaxies, each comparable to our own in size and containing a similar number of stars. The average distance between these galaxies is three million light years. The extent of the visible universe, as it can be seen in the 200-inch telescope, is 20 billion light years.

An analogy will help to clarify the meaning of these enormous distances. Let the sun be the size of an orange; on that scale of sizes the earth is a grain of sand circling in orbit around the sun at a distance of 30 feet; the giant planet Jupiter, 11 times larger than the earth, is a cherry pit revolving at a distance of 200 feet or one city block; Saturn is another cherry pit two blocks from the sun; and Pluto, the outermost planet, is still another sand grain at a distance of ten city blocks from the sun.

On the same scale the average distance between the stars is 2000 miles. The sun's nearest neighbor, a star called Alpha Centauri, is 1300 miles away. In the space between the sun and its neighbors there is nothing but a thin distribution of hydrogen atoms, forming a vacuum far better than any ever achieved on earth. The Galaxy, on this scale, is a cluster of oranges separated by an average distance of 2000 miles, the entire cluster being 20 million miles in diameter.

An orange, a few grains of sand some feet away, and then some cherry pits circling slowly around the orange at a distance of a city block. Two thousand miles away is another orange, perhaps with a few specks of planetary matter circling around it. That is the void of space.

The avg. distance between stars in the Galaxy is 5 light years. The avg. distance between Galaxies is 3 million light years.

I light year = 6 trillion miles

28

THE
HIERARCHY
OF STRUCTURE

The Nucleus and the Atom

Molecules and Solid Matter

The Earth

The Solar System

Our Galaxy

Neighboring Galaxies

Galactic Shapes

Clusters of Galaxies

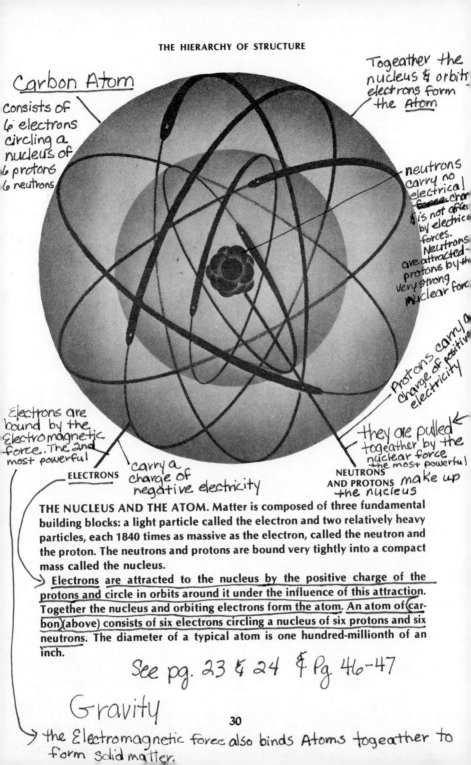

Togeather the nucleus & orbiting electrons form the Atom

Carbon Atom

Consists of 6 electrons circling a nucleus of 6 protons 6 neutrons

neutrons carry no electrical ~~force~~ charge & is not affected by electrical forces. Neutrons are attracted protons by the very strong nuclear force

Protons carry a charge of positive electricity

Electrons are bound by the Electromagnetic force. The 2nd most powerful

they are pulled togeather by the nuclear force the most powerful

ELECTRONS carry a charge of negative electricity

NEUTRONS AND PROTONS make up the nucleus

THE NUCLEUS AND THE ATOM. Matter is composed of three fundamental building blocks: a light particle called the electron and two relatively heavy particles, each 1840 times as massive as the electron, called the neutron and the proton. The neutrons and protons are bound very tightly into a compact mass called the nucleus.

Electrons are attracted to the nucleus by the positive charge of the protons and circle in orbits around it under the influence of this attraction. Together the nucleus and orbiting electrons form the atom. An atom of carbon (above) consists of six electrons circling a nucleus of six protons and six neutrons. The diameter of a typical atom is one hundred-millionth of an inch.

See pg. 23 & 24 & Pg 46-47

Gravity

the Electromagnetic force also binds Atoms togeather to form solid matter.

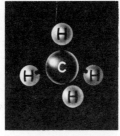
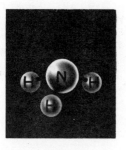

WATER =
2 Hydrogen Atoms
& 1 Oxygen Atom

METHANE =
1 Carbon Atom &
4 Hydrogen Atoms

AMMONIA =
1 Nitrogen Atom &
3 Hydrogen Atoms

MOLECULES AND SOLID MATTER. Atoms are bound together in groups to form molecules. A molecule of water consists of two atoms of hydrogen bound to one atom of oxygen. Other compounds of hydrogen with common elements are methane (a carbon atom bound to four hydrogen atoms) and ammonia (a nitrogen atom bound to three hydrogen atoms).

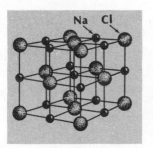
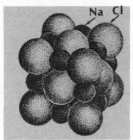

(Na) is sodium
(Cl) is chloride

salt

Large numbers of atoms or molecules joined together form solid matter. Crystals of common salt *(left)* consist of atoms of sodium (Na) and chloride (Cl) arranged in a lattice *(middle)*. <u>One grain of salt contains a million trillion atoms.</u> In the drawing of the lattice, large spaces are shown between the atoms for clarity. In actuality, they are closely packed *(right)*.

 — Hydrogen Atom consists of one electron & one proton. This is the simplest atom. Hydrogen is the most abundant element found in nature. Makes up 90% of all matter

in the Universe

THE EARTH *(below)*. **The earth is a large collection of atoms cemented together in a ball of rock 8000 miles in diameter and weighing 6 billion trillion tons. At the center is a core of nickel and iron 1800 miles in radius. Surrounding the core is a mantle of solid rock 2200 miles thick. The mantle is capped by a crust of lighter rocks whose average thickness is about 15 miles.**

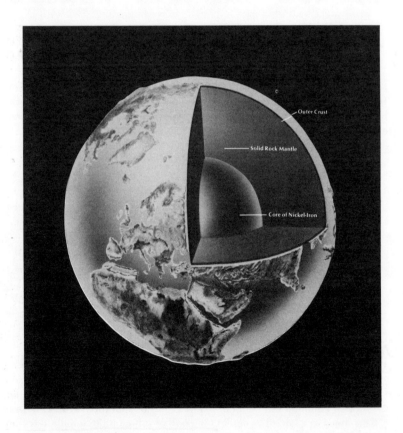

THE SOLAR SYSTEM *(overleaf)*. The earth is one of nine planets that revolve in orbits around the sun, bound to it by the force of gravity. The weight of the sun is 700 times greater than the combined weight of the nine planets. The (sun) the (planets) their (thirty-four satellites) and a large number of other lesser bodies, including asteroids and comets, form the solar system.

The orbits of all the planets are within a few degrees of one plane, except for the orbits of the innermost planet, (Mercury) and the outermost planet, (Pluto) All orbits are close to perfect circles, with the exception, again, of (Mercury) and (Pluto) Pluto circles the sun at a distance varying from 3 billion to 4 billion miles. The orbit of Pluto marks the outer boundary of the solar system.

In the space between the planets, and outside the boundary of the solar system in the space between the stars, there is a tenuous cloud of hydrogen gas with a density of 10 to 100 atoms per cubic inch. The distance from the sun and its planets to the nearest known star, Alpha Centauri, is 25 trillion miles, or 5000 times the size of the solar system. Outer space, like the inner space of the atom, is nearly empty.

Our solar system is our sun & planets. There are many solar systems within our galaxy. There are many galaxies within our Universe.

NEPTUNE

URANUS

THE ~~SOLAR~~ *Our* SOLAR SYSTEM

PLUTO

MERCURY

VENUS

SUN

MARS

ITER

MOON

EARTH

ASTEROID BELT

SATURN

OUR GALAXY. The sun is one of 200 billion stars bound together by the force of gravity into a large cluster called the Galaxy. Only a few thousand of these stars can be seen with the naked eye. However, many more appear in photographs taken with a large telescope *(opposite)*. Approximately 10,000 stars are visible in this photograph, although it represents only a thousandth of the area of the full night sky. The photograph was obtained with a 10-inch telescope by an exposure of several hours.

The Galaxy is flattened by its spinning motion into the shape of a disk, with the sun located three-fifths of the way from the center of the disk to the edge. The artist's diagram *(below)* shows the structure of the Galaxy, viewed edge-on, with the position of the sun indicated.

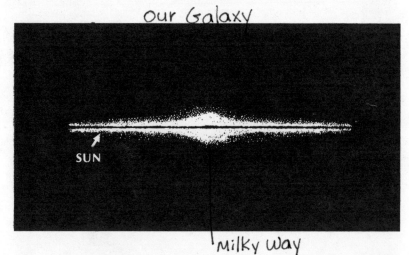

Our Galaxy

SUN

Milky Way

The distances between stars are usually expressed in terms of the light year, which is the distance covered in one year by a ray of light traveling at 186,000 miles per second. One light year is 6 trillion miles. The average distance between stars in our part of the Galaxy is 6 light years. The diameter of the entire Galaxy is 100,000 light years, and the thickness of the central disk is about 2000 light years.

The concentration of stars in the central disk of the Galaxy makes it an exceedingly bright region when viewed edge-on. On a clear evening, if we are away from city lights, we see this band of light stretching across the sky. It is called the Milky Way.

A montage of photographs displaying the Milky Way appears on the following pages.

STARS IN OUR GALAXY

THE MILKY WAY: AN EDGE-ON VIEW OF OUR GALAXY. This montage of photographs shows the Galaxy viewed edge-on, as it would appear to an observer in our solar system. The 7000 brightest stars have been drawn in separately. The bright, luminous clouds consist of billions of individual

stars. The dark regions running irregularly through the center of the Milky Way are clouds of dust which absorb the light coming to our solar system from distant stars in the Galaxy.

NEIGHBORING GALAXIES. Our nearest extragalactic neighbors are two small galaxies, satellites of our own, which are held captive by the gravitational force of the 200 billion stars in our galaxy. Each of these satellite galaxies contains a few billion stars. They are visible to the naked eye as faint patches of light, called the Magellanic Clouds.

Roughly a dozen other galaxies exist within 3 million light years of our galaxy. Their positions are shown below, together with the titles assigned to them in astronomical catalogues.

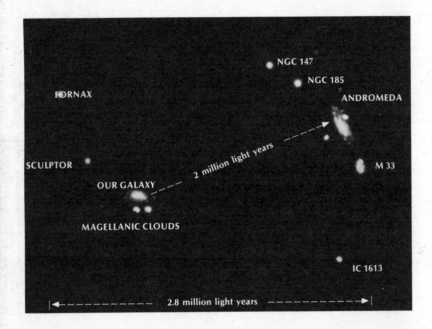

The nearest galaxy of comparable size to our own is the Great Nebula in Andromeda, located approximately 2 million light years from us. The Andromeda galaxy closely resembles ours in size, shape and number of stars. At the right is a photograph of the Andromeda galaxy taken with the 200-inch telescope on Mount Palomar. The small points of light are individual stars within our own galaxy, lying in the line of sight to Andromeda. The two luminous regions above and below Andromeda are satellite galaxies, similar to the Magellanic Clouds.

NGC 205
see next p1.

The Andromeda galaxy

GALACTIC SHAPES. The shape of our galaxy is illustrated clearly in the sequence opposite, which shows four different disk-shaped galaxies, all resembling ours but tilted at various angles to our line of sight. The galaxies are identified by the catalogue numbers assigned to them by astronomers. The fourth photograph shows the spiral arms characteristic of the type of galaxy to which the sun belongs. These spiral arms are concentrations of gas and dust in which stars are born.

Not all galaxies have this shape. Some are irregular; others are round or elliptical like NGC 205 *(below)*. This galaxy, also visible at the lower left on page 41, is a small satellite of the Andromeda galaxy, and contains about four billion stars. Most are very old, having been created in a single burst of star formation when NGC 205 was born, nearly 20 billion years ago. Stars such as these, formed early in the history of the universe, consist of primordial hydrogen and helium. They do not contain the heavier elements which are products of supernova explosions, such as carbon, oxygen, silicon and iron. Therefore solar systems in NGC 205 and similar galaxies cannot contain planets like the earth, with the chemical ingredients of life.

NGC 205

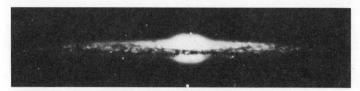

NGC 4565   Viewed edge-on

NGC 4216   Tilted 15 degrees

NGC 7331   Tilted 30 degrees

NGC 5457   Viewed face-on

CLUSTERS OF GALAXIES. Ten billion other galaxies are within range of the 200-inch telescope on Mount Palomar. Most of these galaxies occur in clusters, ranging from as few as 3 to as many as 10,000 galaxies in one cluster. The galaxies in a cluster are separated by an average distance of 1 million light years. Our galaxy is a member of a small cluster of galaxies called the Local Group, which includes the large galaxy in Andromeda and approximately a dozen others. Several galaxies in the Local Group are shown in the drawing on page 40.

The photograph opposite shows a group of approximately 50 galaxies at the center of the giant cluster of galaxies in the Constellation Hercules, about 300 million light years distant. The Hercules cluster, containing upward of 10,000 galaxies, is one of the major systems of organized matter in the universe. In the photograph, the spiked objects and some of the perfectly circular spots are individual stars in our galaxy; every other object is a galaxy in the Hercules cluster.

Clusters of galaxies are the largest known systems of organized matter. They terminate the hierarchy of structure in the universe.

A CLUSTER OF GALAXIES ·

All objects are held togeather by these 3 forces:
1) *nuclear force*
2) *the force of electromagnetism*
3) *the force of gravity*

there powers are in order. 1) being the strongest.

3 THE FORCES OF NATURE

> There are therefore Agents in Nature able to make the Particles of Bodies stick together by very strong Attractions. And it is the Business of experimental Philosophy to find them out.
>
> —ISAAC NEWTON, *Opticks* (1704)

It is remarkable that all objects in the universe, from the smallest nucleus to the largest galaxy, are held together by only three fundamental forces—a nuclear force, the force of electromagnetism, and the force of gravity.

Most powerful is the nuclear force,* which binds neutrons and protons together into the nucleus of the atom. This extremely strong force of attraction pulls the particles of the nucleus together into an exceedingly compact body with a density of one billion tons per cubic inch.

Next strongest is the electrical (electromagnetic) force, which is approximately 100 times weaker than the nuclear force. This force binds the electrons to the nucleus to form atoms, and binds atoms together into solid matter.

Least powerful is the force of gravity. The gravitational force is extremely weak, about 10^{38} times weaker than the nuclear force, and 10^{36} times weaker than the force of electricity.[†] These numbers are

---

*Two kinds of nuclear force exist—the one discussed here, and a weaker one, called the weak force. The latter plays no substantial role in this discussion, and I omit it for simplicity.

† 10^{38} is 1 followed by 38 zeros; one says: "10 to the 38th." It is a convenient shorthand for writing very large numbers. Some of these numbers have names. For example, 10^3 (= 1000) is one thousand; 10^6 = 1 million; 10^9 = 1 billion; 10^{12} = 1 trillion = 1 million million. Numbers larger than this, however, do not have names in common usage.

exceedingly large; 10^{36} is a trillion times larger than the number of grains of sand in the oceans of the earth. Nonetheless it is this very frail agent which keeps the moon in orbit around the earth, the earth and other planets revolving around the sun, and the sun and other stars clustered together in our galaxy.

In one sense gravity is the simplest of the basic forces. Its simplicity lies in the fact that the force of gravity acting between two objects always pulls them together and never pushes them apart. The electrical force is more complicated because its effect on some pairs of particles is to pull them together, while on others its effect is to push them apart. The explanation of these two kinds of action has come out of laboratory experiments with electricity during the last 200 years. These experiments show that two types of electricity exist, called positive and negative. When two particles bear the same kind of charge, i.e., both negative or both positive, they repel each other; but two particles bearing different kinds of charge are attracted to one another.

Every electron carries a charge of negative electricity; therefore all electrons repel one another. Similarly, every proton carries a charge of positive electricity; therefore all protons repel one another. However, the electron and proton attract one another because they carry different kinds of electricity.

This last fact enables us to understand the existence of the atom. The nuclear force acts as a glue which binds the neutrons and protons together to form the nucleus. The protons within the nucleus, being positively charged, exert an electrical attraction on any electrons that may be in the neighborhood. Under the influence of this attraction electrons are captured by the nucleus and forced to circle around it, as the planets circle the sun under the attraction of gravity. The nucleus and its surrounding electrons together constitute the atom.

The simplest atom consists of a single electron circling around a nucleus composed of a single proton. This is the atom of hydrogen, the most abundant element found in nature, which makes up 90 percent of all matter in the universe.

The neutron is the other basic nuclear particle; it carries no electric charge, and is not affected by electrical forces. However, neutrons are attracted to protons by the very strong nuclear force. Under the influence of this attraction, a neutron and a proton may be joined into a single, heavier nucleus, weighing twice as much as the proton. The new nucleus contains one positive charge, and therefore attracts a single electron to itself, forming an atom similar to hydrogen, but twice as massive. This atom is called heavy hydrogen, or deuterium. Heavy hydrogen was discovered by Harold Urey in 1932, an achievement for which he received the Nobel Prize in 1934. It enters into combination with oxygen to form heavy water, a substance identical with water but somewhat denser. Heavy water is a relatively scarce substance, there being one molecule of heavy water to each 10,000 molecules of ordinary water in the oceans.

After hydrogen, the second most abundant element in the universe is helium. Helium usually exists on the earth in the form of a gas. It is lighter than air, once had a wide use in dirigibles, and is still used in children's balloons. The nucleus of a helium atom contains two neutrons and two protons. Since each proton can attract one electron to itself, an atom of helium has two electrons circling in orbit around the nucleus.

Hydrogen and helium together constitute approximately 99 percent of the matter in the universe. All other elements make up the remaining one percent. Among the elements in this last one percent, the most abundant is the critically important substance, oxygen. An atom of oxygen is composed of a nucleus containing eight neutrons and eight protons, around which eight electrons circle. Other nuclei exist with still greater numbers of neutrons and protons. In each case, the complete atom is formed when the nucleus has circling around it a number of electrons equal to the number of protons it contains. In the heaviest elements, such as lead and gold, the nucleus contains approximately 200 neutrons and protons, surrounded by up to 92 electrons arranged in a complicated array of orbits of many different sizes. The atom of uranium is the largest, heaviest and most complicated of all. It consists of a nucleus containing 146 neutrons and 92 protons, surrounded by 92 electrons.

Altogether there exist 92 different kinds of nuclei and 92 corresponding kinds of atoms, stretching from hydrogen to uranium. These elements, in various combinations, make up all the varieties of matter—animal, vegetable and mineral—which exist in the universe.

The most abundant ~~form~~ element

Hydrogen atom =
1 proton & 1 electron

Heavy hydrogen atom (deuterium) =
1 proton & 1 neutron & 1 electron

2nd most abundant element

Helium atom =
2 neutrons & 2 protons & 2 electrons

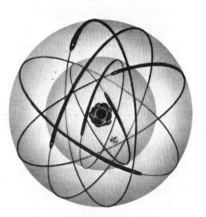

Carbon atom

SIMPLE ATOMS. The diagram shows atoms of hydrogen, heavy hydrogen, helium and carbon. Each atom consists of a nucleus composed of neutrons and protons, surrounded by electrons which orbit the nucleus at a considerable distance. The protons carry a positive electric charge which attracts negatively charged electrons until the number of electrons orbiting around the nucleus is equal to the number of protons it contains. Each hydrogen atom contains one electron, which orbits the single proton in the nucleus; the helium atom contains two orbiting electrons and two nuclear protons; the carbon atom contains two electrons in an inner orbit and four electrons in an outer orbit, or six in all, balancing the six protons in the nucleus.

Most properties of an element depend on the number of electrons in the atom; this number is in turn fixed by the number of protons in the nucleus. Thus the nucleus ultimately determines the properties of all elements.

The largest & heaviest & most complicated of all Atoms is uranium. [49] Consisting of 146 neutrons, 92 protons & 92 electrons.

4   THE ALCHEMIST'S GOAL

The Ancients believed that all the varieties of matter in the world could be made by mixing together the four basic substances—earth, air, fire and water—in the right proportions and with suitable amounts of dryness, moisture, heat and cold. A strong element of mysticism often pervaded the experiments. Correct proportions were not enough; suitable incantations and recipes had to be followed. These beliefs nourished the science of alchemy throughout the Middle Ages down to the time of Isaac Newton.

Newton himself was an enthusiastic alchemist. His nephew and amanuensis, Humphrey Newton, wrote, "About six weeks in the spring and six weeks in the fall, the fire in the laboratory scarcely went out . . . . he would sometimes look into an old moldy book which lay in his laboratory. I think it was called *Agricola de Metallis*, the transmuting of metals being his chief design . . . ." Newton appears to have taken greater pleasure in these alchemical experiments than in the labors in mathematics and physics which secured his fame. He believed in the transmutation of elements, specifically of lead into gold, and was interested in the money-making possibilities of such endeavors. In 1669 he wrote in a letter to his friend Francis Aston, "And if you meet with any transmutations out of their own species into another . . . those, above all, will be worth your noting, being the most *luciferous,* and many times *lucriferous* experiments in philosophy." "Lucriferous" was a term in usage among alchemists to describe the potentially profitable applications of their science, as opposed to the luciferous applications — those which added to the sum of knowledge.

Newton's exceptional talents led him to the invention of the calculus and the discovery of the law of universal gravity. Yet the architect of the universe failed completely in his effort to transmute the elements. The reasons for his failure could not have been perceived by anyone in the seventeenth century. They did not become apparent until Rutherford's discoveries in the first years of the twentieth century. Then it became clear that Newton and the alchemists had been tampering only with the superficial qualities of matter. They were not striking at the heart of the problem—the nucleus.

It is in the nucleus that the quintessence of a substance resides. What is the difference between lead and gold? Each is a heavy metal; one is base, common and of a dull gray color, the other noble, rare and a lustrous yellow. Yet the two elements are surprisingly close in the hierarchy of atomic structure. An atom of gold is composed of a nucleus containing 118 neutrons and 79 protons, around which 79 electrons circle in orbit. An atom of lead contains 126 neutrons and 82 protons surrounded by 82 electrons. The eight additional neutrons in the lead nucleus serve only to make it a somewhat heavier element than gold; but the three additional protons play a more important role, for they draw three extra electrons to the lead atom. It is these extra electrons which are felt, seen and touched; they are responsible for the difference between lead and gold. Yet the number of electrons in an atom is controlled entirely by the number of protons in the nucleus. If three of the protons in the nucleus of an atom of lead could be removed in some way, the resulting nucleus, with 79 protons, would be the nucleus of an atom of gold.

But it is not possible to dislodge protons from the nucleus very easily. The fires of the alchemist's furnace can do no more than melt a bar of lead by breaking the bonds between neighboring atoms. Perhaps, if the flame is hot enough, some of the atoms may lose an electron. But an atom of lead stripped of one electron is still an atom of lead. It still contains 82 protons, and if it lacks an electron it will exert an electrical attraction on other electrons nearby until it has collected the full complement of 82 to become a normal atom of lead. The nature of an element can be changed only by adding protons to the nucleus or removing them. But the force that keeps the proton within the nucleus is far stronger than the force that binds the electron to the atom or the force that cements atoms together into solid matter. Because the proton is locked so firmly within the nucleus, an

Gold
118 neutrons 79 electrons
79 protons

Lead
126 neutrons 82 electrons
82 protons

exceptional degree of force must be applied to dislodge it. An ordinary blow with a hammer will not do, because the hammer cannot smash through the electrons of the atom to reach the nucleus at the center. Only an encounter with another minute particle, moving at speeds great enough to penetrate the electron barrier around the nucleus, will suffice.

Such an encounter was observed for the first time by Rutherford in a classic experiment carried out in 1919, in which he bombarded nitrogen atoms with high-speed helium nuclei emitted from radium. In a few cases, a fast-moving helium nucleus collided directly with a nitrogen nucleus and stuck to it to form a heavier particle, which was the nucleus of a new element.

What new element was created by Rutherford? The nitrogen nucleus contains seven neutrons and seven protons; the helium nucleus contains two neutrons and two protons. The nucleus resulting from their union therefore contains nine neutrons and nine protons. Rutherford found that one of the protons was dislodged by the violence of the impact and left the scene of the collision, leaving a residual nucleus with eight protons. The nucleus then collected eight electrons in orbit around it to become an atom of *oxygen*. This atom, containing nine neutrons, was slightly heavier than the common variety of oxygen, whose nucleus contains only eight neutrons, but it was nonetheless, a bona fide oxygen atom. The experiment had transmuted nitrogen into a different substance, oxygen.

Of course the yield of the newly created element was exceedingly small. Only one helium nucleus in 50,000 collided with a nitrogen nucleus and fused with it to create oxygen; hence, very few oxygen atoms resulted from the experiment. Nevertheless, Rutherford's experiment was a breakthrough, for it showed the way in which new elements can be created by building up large nuclei out of smaller ones. We suspect today that all the elements in the sun and the planets, and in our bodies, were created in the centers of other stars, earlier in the history of the Galaxy, by the same kind of nuclear bombardment which Rutherford performed in the laboratory.

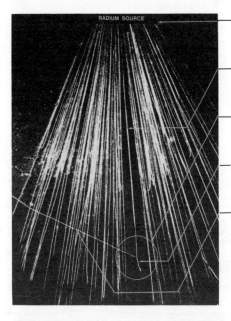

RADIUM SOURCE

1. Tracks of helium nuclei emerge from radium source at top.

2. Arrow points to track of helium nucleus which will undergo collision below.

3. Circle marks site of collision between helium nucleus and nitrogen nucleus in cloud chamber.

4. Short track moving downward to right is produced by nucleus recoiling from collision.

5. Proton knocked out during collision moves upward to left.

(Some helium tracks have been removed to exhibit the track of the nucleus involved in the collision more clearly.)

RUTHERFORD'S TRANSMUTATION OF AN ELEMENT. In order to change the basic properties of an element, it is necessary to change its nucleus. The photograph above was obtained in one of the first experiments performed by Rutherford in his Cambridge laboratory in which one element was artificially transformed into another. The photograph shows a cloud chamber — a vessel filled with water vapor and illuminated from the side to show the tracks of particles passing through it. A thin foil of radium, shown schematically at the top, emits fast-moving helium nuclei at a steady rate as the products of a radioactive decay process. Some fifty tracks produced by these helium nuclei are visible, spreading downward through the chamber from the radium foil (1).

The cloud chamber contains nitrogen gas in addition to water vapor. Among the many helium nuclei which pass through the chamber, one (2) collides with the nucleus of an atom of nitrogen and fuses with it to form a new nucleus. The collision occurs at the center of the circle (3). The resultant nucleus is that of oxygen; its track may be seen as a short stub moving downward and slightly to the right within the circle (4). The force of the impact has dislodged a proton, whose track is visible moving to the left and upward (5).

The artificial transmutation of nitrogen into oxygen by Rutherford, as demonstrated in this cloud-chamber photograph, marked the first time that man achieved the alchemist's goal by changing one element into another.

5 RED GIANTS, WHITE DWARFS AND BLACK HOLES IN SPACE

The stars seem immutable, but they are not. They are born, evolve and die like living organisms. The life story of a star begins with the simplest and most abundant element in nature, which is hydrogen. The universe is filled with thin clouds of hydrogen, which surge and eddy in the space between the stars. In the swirling motions of these tenuous clouds, atoms sometimes come together to form small pockets of gas. These pockets are temporary condensations in an otherwise highly rarefied medium. Normally the atoms fly apart again in a short time as a consequence of their random motions, and the pocket of gas quickly disperses to space. However, each atom exerts a small gravitational attraction on its neighbor, which counters the tendency of the atoms to fly apart. If the number of atoms in the pocket of gas is large enough, the accumulation of all these separate forces will hold it together indefinitely. It is then an independent cloud of gas, preserved by the attraction of each atom in the cloud to its neighbor.

With the passage of time, the continuing influence of gravity, pulling all the atoms closer together, causes the cloud to contract. The individual atoms "fall" toward the center of the cloud under the force of gravity; as they fall, they pick up speed and their energy increases. The increase in energy (heats) the gas and raises its temperature. The shrinking, continuously self-heating ball of gas is an embryonic star.

As the gas cloud contracts under the pressure of its own weight, the temperature at the center mounts steadily. When it reaches 100,000 degrees Fahrenheit, the hydrogen atoms in the gas collide with sufficient violence to dislodge all electrons from their orbits around the protons. The original gas of hydrogen atoms, each consisting of an electron circling around a proton, becomes a mixture of two gases, one composed of electrons and the other of protons.

At this stage the globe of gas has contracted from its original size, which was 10 trillion miles in diameter, to a diameter of 100 million miles. To understand the extent of the contraction, imagine the Hindenburg dirigible shrinking to the size of a grain of sand.

The huge ball of gas—now composed of separate protons and electrons—continues to contract under the force of its own weight, and the temperature at the center rises further. After 10 million years the temperature has risen to the critical value of 20 million degrees Fahrenheit.* At this time, the diameter of the ball has shrunk to one million miles, which is the size of our sun and other typical stars.

Why is 20 million degrees a critical temperature? The explanation is connected with the forces between the protons in the contracting cloud. When two protons are separated by large distances, they repel one another electrically because each proton carries a positive electric charge. But if the protons approach within a very close distance of each other, the electrical repulsion gives way to the even stronger force of nuclear attraction. The protons must be closer together than one 10-trillionth of an inch for the nuclear force to be effective. Under ordinary circumstances, the electrical repulsion serves as a barrier to prevent as close an approach as this. In a collision of exceptional violence, however, the protons may pierce the electrical barrier which separates them, and come within the range of their nuclear attraction. Collisions of the required degree of violence first begin to occur when the temperature of the gas reaches 20 million degrees.

Once the barrier between two protons is pierced in a collision, they pick up speed as a result of their nuclear attraction and rush rapidly toward each other. In the final moment of the collision the force of nuclear attraction is so strong that it fuses the protons together into a

---

*Twenty million degrees is a very high temperature. For comparison, the temperature of the flame in the gas burner of the kitchen stove is 1000 degrees, and the temperature of the hottest steel furnace is 10,000 degrees.

single nucleus. At the same time the energy of their collision is released in the form of heat and light. This release of energy marks the
birth of the star.

The energy passes to the surface and is radiated away in the form
of light, by which we see the star in the sky. The energy release,
which is one million times greater per pound than that produced in a
TNT explosion, halts the further contraction of the star, which lives
out the rest of its life in a balance between the outward pressures
generated by the release of nuclear energy at its center and the inward
pressures created by the force of gravity.

The fusion of two protons into a single nucleus is only the first step
in a series of reactions by which nuclear energy is released during the
life of the star. In subsequent collisions, two additional protons are
joined to the first two to form a nucleus containing four particles.
Two of the protons shed their positive charges to become neutrons in
the course of the process. The result is a nucleus with two protons
and two neutrons. This is the nucleus of the helium atom. Thus, the
sequence of reactions transforms protons, or hydrogen nuclei, into
helium. *

The fusion of hydrogen into helium is the first and longest stage in
the history of a star, occupying about 90 percent of its lifetime.
Throughout this long period of the star's life its appearance changes
very little, but toward the end of the hydrogen-burning stage, when
most of the hydrogen has been converted into helium, the star begins
to show the first signs of age. The telltale symptoms are a swelling
and reddening of the outer layers, commencing imperceptibly and
progressing until the star has grown to a huge red ball 100 times

---

* The transmutation of heavy hydrogen into helium and heavier elements has been
duplicated on the earth for brief moments in the explosion of the hydrogen bomb.
However, we have never succeeded in fusing hydrogen nuclei under controlled conditions in such a way that the energy released can be harnessed for constructive pur-
poses. The United States, the Soviet Union and other countries have invested pro-
digious amounts of money and energy in the effort, for the stakes are high, but
physics has not yet been equal to the task. The difficulty is that no furnace has yet
been constructed on the earth whose walls can contain a fire at the temperature of
the millions of degrees necessary to produce nuclear fusion. The only furnace that
can do this is provided by nature in the heart of a star.

larger than its original size. The sun will reach this stage in another 6 billion years, at which time it will have swollen into a vast sphere of gas engulfing the planets Mercury and Venus and reaching out nearly to the orbit of the earth. This red globe will cover most of the sky when viewed from our planet. Unfortunately we will not be able to linger and observe the magnificent sight, because the rays of the swollen sun will heat the surface of the earth to 4000 degrees Fahrenheit and eventually evaporate its substance. Perhaps one of the moons of the outer planets will be a suitable habitat for us by then. More likely, we will have fled to another part of the Galaxy. Such distended, reddish stars are called *red giants* by astronomers. An example of a red giant is Betelgeuse, a fairly bright star in the constellation Orion, which appears distinctly red to the naked eye.

A star continues to live as a red giant until its reserves of hydrogen fuel are exhausted. With its fuel gone, the red giant can no longer generate the pressures needed to maintain itself against the crushing inward force of its own gravity, and the outer layers begin to fall in toward the center. The red giant collapses.

At the center of the collapsing star is a core of pure helium, produced by the fusion of hydrogen throughout the star's earlier existence. Helium does not fuse into heavier nuclei at the ordinary stellar temperature of 20 million degrees, because the helium nucleus, with *two* protons, carries a double charge of positive electricity, and, as a consequence, the electrical repulsion between two helium nuclei is stronger than the repulsion between two protons. A temperature of 200 million degrees is required to produce collisions sufficiently violent to pierce the electrical barrier between helium nuclei.

As the star collapses, however, heat is liberated and its temperature rises. Eventually the temperature at the center reaches the critical value of 200 million degrees. At that point helium nuclei commence to fuse in groups of three to form carbon nuclei, releasing nuclear energy in the process and rekindling the fire at the center of the star. The additional release of energy halts the gravitational collapse of the star. It has obtained a new lease on life by burning helium nuclei to produce carbon.

In stars the size of the sun, the helium-burning stage lasts for about one hundred million years. At the end of that time the reserves of fuel, composed now of helium rather than hydrogen, once again are exhausted, and the center of the star is filled with a residue of carbon nuclei. These nuclei, possessing *six* positive electrical charges, are separated by an even more formidable electrical barrier than helium nuclei, and collisions of even greater violence are required for its penetration. The 200-million-degree temperatures which fuse helium nuclei are not adequate for the fusion of carbon nuclei; no less than 600 million degrees are required.

Since the temperatures prevailing within the red giant fall short of 600 million degrees, the nuclear fires die down as the carbon accumulates, and the star, once again lacking the resources needed to sustain it against the weight of its outer layers, commences to collapse a second time under the force of gravity.

All stars lead similar lives up to this point, but their subsequent evolution and manner of dying depend on their size and mass. The small stars shrivel up and fade away, while the large ones disappear in a gigantic explosion. The sun lies well below the dividing line; we are certain that it will fade away at the end of its life.

The paths of the small and large stars diverge because of differences in the amount of heat generated during the second collapse, at the end of the red giant stage. In a small star the collapse generates a modest amount of heat, and the temperature at the center fails to reach the 600-million-degree level required for the ignition of carbon nuclei. Thus the nuclear fire is never rekindled. Instead the collapse continues, until finally the matter within the star is so compressed that it resists a further reduction in size. The star then remains forever in this highly compressed state. Roughly the size of the earth, it has been squeezed by the force of its own weight into a space only a millionth as large as the volume it originally occupied. A teaspoon of matter from the center of this compact body would weigh 10 tons. If we should ever come across such a star, even though its surface temperature might have declined to a comfortable level, we would be unable to land on this strange world, because its gravity would crush

2nd Collapas is when the helium is now carbon

white dwarf

a visitor with a force of 100 million pounds.

Although the center of the star never gets hot enough to burn carbon, the temperature of the surface rises sufficiently so that the star appears white-hot to the eye. These shrunken white-hot stars are called *white dwarfs*. Slowly the white dwarf radiates into space the last of its heat. In the end its temperature drops, and it fades into a blackened corpse.

A very different fate awaits a large, massive star. Because the weight of the star is so great, its collapse generates an enormous amount of heat, greater than the heat generated in the creation of the white dwarf. Soon the temperature reaches the critical level of 600 million degrees at which carbon nuclei fuse together. The fusion of the carbon nuclei forms still heavier elements ranging from oxygen up to sodium.

Eventually, the carbon fuel reserves are also exhausted; once again their exhaustion is followed by further stages of collapse, heating, and renewed nuclear burning, leading to the production of still other elements.

In this way, through the alternation of collapse and nuclear burning, a massive star successively manufactures all elements up to iron. But iron is a very special element. This metal, which lies halfway between the lightest and heaviest elements, has an exceptionally compact nucleus, whose neutrons and protons are so tightly packed that no energy can be squeezed from it in any sort of nuclear reaction. In fact, nuclear reactions in iron absorb energy; they have the same effect as water thrown on hot coals. When a large amount of iron accumulates at the center of the star, the fire cannot be rekindled; it goes out for the last time, and the star commences a final collapse under the force of its own weight.

The ultimate collapse is a catastrophic event. The iron nuclei at the center soak up the energy of the star as fast as it is generated, and the collapsing materials, meeting negligible resistance, fall in toward the center at enormous speeds, covering a million miles in less than a minute. They pile up around the material at the center in a dense lump, squeezing the core of the star, and creating enormous pressures. When the pressure at the center is sufficently great, the collapse comes to a halt. The collapsed star, compressed like a spring, is momentarily still—and then it rebounds in a violent explosion,* leaving nothing behind but the squeezed remnant of the star's original core.

The explosion generates temperatures ranging up to trillions of degrees. At these temperatures some of the nuclei in the exploding star are disintegrated, and many neutrons are freed. The neutrons are captured by other nuclei, building up the heavier elements, such as lead, gold and uranium. In this way the remaining elements of the periodic table, extending beyond iron, are manufactured in the final moments of the star's life. This is why these heavy elements, including silver, gold, platinum, and so on, are so rare in the universe and so rare on the earth. The small amounts that exist in the universe were manufactured in minutes or even seconds, in the last gasp of a star's life. Elements like carbon and oxygen are more abundant because they were manufactured steadily within the star over the course of many millions of years.

The explosion of the star hurls out to space all the elements it has been manufacturing during its lifetime. The entire episode lasts a few minutes, from the onset of the collapse to the final explosion. This is a short interval for the demise of an object which may have lived for ten million years.

The exploding star is called a supernova. Supernovas blaze up with a brilliance many billions of times greater than the brightness of the sun; if the supernova happens to be nearby in our Galaxy, it appears suddenly as a new star in the sky, brighter than any other, and easily visible with the naked eye in the daytime. The last supernova seen in Europe exploded in 1604 and caused a sensation. One of the earliest reported supernovas was a brilliant explosion recorded by Chinese astronomers in A.D. 1054. At the position of this supernova there is

---

* Sometimes the star blows up as soon as its carbon ignites, without going through the sequence of events leading up to the formation of iron. The end result is the same: a violent explosion occurs; the star is destroyed; and its materials are dispersed to space.

today a great cloud of gas known as the Crab Nebula, expanding outward at a speed of seven hundred miles per second, which contains the remains of the star that exploded nine hundred years ago.

What happens to the compressed core of the supernova after its outer layers explode into space? The answer to this question was unknown until 1967. In that year, pulsars—the most interesting objects to be found in the sky for many years—were discovered.

The discovery came about by pure chance. Jocelyn Bell, an astronomy student at Cambridge University, was assigned the task of investigating fluctuations in the strength of radio waves from distant galaxies. She found unexpectedly that certain places in the heavens were emitting short, rapid bursts of radio waves at regular intervals. Each burst lasted no more than one hundredth of a second. The rapid succession of bursts seemed like a speeded-up, celestial Morse code.

The interval between successive bursts was extraordinarily constant. In fact, it did not change by more than one part in 10 million. A clock with this precision would gain or lose no more than a second a year.

No star or galaxy had ever been observed to emit signals as bizarre as these. At first, some astronomers thought that intelligent beings on other stars might be beaming a message to the earth, and referred to the Morse-code stars as LGM's, standing for Little Green Men. But it soon became evident that the radio pulses had a natural and not an artificial origin. One of the main reasons for this conclusion was the fact that the signals were spread over a broad band of frequencies. If an extraterrestrial society were trying to signal other solar systems, its interstellar transmitters would require enormous power in order for the signals to carry across the trillions of miles that separate every star from its neighbors. The only feasible way to do this would be to concentrate all available power at one frequency, as we do when we broadcast radio and TV programs. It would be wasteful, purposeless and unintelligent to diffuse the power of the transmitter over a broad band of frequencies.

This cold reasoning dashed the hopes of romantics who believed for a short time that man might be receiving his first message from outer space. "LGM's" disappeared from scientific conversation, "pulsars" took their place, and scientists settled down to search for a natural explanation of the peculiar signals.

The first clue to the answer was the sharpness of the pulses. From the fact that each pulse lasted for one hundredth of a second or less, astronomers concluded that a pulsar is an incredibly small object for a star, far smaller than a white dwarf. This conclusion was based on the fact that when an object emits a burst of radio waves, the waves from different parts of the object arrive at the earth at different times, blurring the sharpness of the original pulse. The smaller the object, the sharper the pulse. Following this line of reasoning, the astronomers calculated that the objects were no more than 10 miles in radius.

This is a startling conclusion. Until then the white dwarf—about 10,000 miles in radius—was thought to be the smallest, densest star in the universe. How could a star be a thousand times smaller than a white dwarf? The matter in such a star would be a billion times denser than the matter in a white dwarf. If the entire earth were compressed to the same degree as a pulsar, it would fit into the Pentagon. If the Pentagon were compressed as much, it would be the size of a pinhead.

The answer goes back to a prediction made several decades ago. At that time, several theoretical astronomers pointed out that when a large star collapses at the end of its life, just before it explodes as a supernova, the materials of the collapsing star pile up at the center, and produce enormous pressures, even more powerful than the inward pressure produced by the star's own weight. Under this crushing burden, the individual electrons and protons within the star are forced to combine into neutrons. A pure ball of neutrons forms at the center of the star, only 10 miles in radius, but with most of the star's original mass packed into it. The hypothetical ball of neutrons was dubbed a "neutron star."

Starting in 1965, astronomers searched for neutron stars assiduously, investigating with particular care the region at the center of the Crab Nebula, where the squeezed-down core of the supernova explosion of A.D. 1054 should have been located. But no neutron stars were discovered, and interest in them faded.

In 1968 a wave of excitement spread through the astronomical community when a pulsar was discovered at the center of the Crab Nebula, at precisely the place where they had previously searched for a neutron star. Suddenly, many items of evidence fitted together like the pieces of a jigsaw puzzle: a neutron star was predicted to exist at the center of the Crab Nebula; a pulsar was discovered at the center of the Crab Nebula; and the neutron star and the pulsar are the only objects known that have the mass of a star packed into a 10-mile sphere. Clearly, neutron star and pulsar were two names for the same thing: a fantastically compressed, super-dense ball of matter, created when a massive star collapses at the end of its life.

One mystery remained to be explained. What produces the sharp, regularly repeated bursts of radiation from which pulsars derive their name? The answer is believed to be that a pulsar, like the sun and most other stars, is subject to violent storms which may last for years, spraying particles and radiation out into space. Each storm occurs in a localized area on the surface of the pulsar and sprays its radiation into space in a narrowly defined direction. When the earth lies in the path of one of these streams of radiation, our radio telescopes pick up the signals, which indicate to us the presence of the pulsar.

But if the spray of radiation is emitted steadily from the pulsar, why do we observe it as a succession of isolated sharp bursts? The reason is probably that pulsars, like most stars, spin on their axes. In fact, being smaller than normal stars, pulsars can spin very rapidly, as many as several times a second. As the pulsar spins, the stream of radiation from its surface sweeps through space like the light from a revolving lighthouse beacon. If the earth happens to lie in the path of the rotating beam, it will receive a sharp burst of radiation once in every turn of the pulsar.

This theory can be checked, because all spinning objects slow down steadily in the course of time as a result of friction. Thus the interval of time between successive bursts of radiation from the pulsar must increase. In 1969 this prediction was confirmed by the discovery that the time between successive pulses from the Crab Nebula pulsar was getting longer, at the tiny but measurable rate of one-billionth of a second per day.

With the realization of the connection between neutron stars, pulsars and supernovas, many astronomers feel that the story of the birth and death of stars is complete. But others suspect that at least one surprise is still in store for us, for there is reason to believe that the neutron star or pulsar is not the ultimate state of compression of stellar matter. If the star is exceedingly massive, the collapse at the end of its life can squeeze the star's core down to a size even smaller than that of a neutron star. When the core is compressed to a radius of two miles, the theory of relativity predicts the occurrence of an extraordinary phenomenon.

According to Einstein's theory, energy and mass are equivalent. The equivalence is expressed in the famous equation

$$E = mc^2$$

where E is the energy of the light ray, m is its mass and c is the velocity of light. Reference is frequently made to this equation in calculating the energy E yielded by the annihilation of an amount m of uranium in the explosion of a nuclear bomb. What has this to do with the collapsing star? If energy is equivalent to mass, a ray of light which possesses electromagnetic energy must also possess mass, just as if it were a particle of matter. Thus, a ray of light emitted from a star will be pulled back by the star's gravity, as a ball thrown up from the surface of the earth is pulled back by the earth's gravity. When the star is normal in size—about a million miles in diameter—the force of gravity on its surface is not strong enough to keep the light rays from escaping, and they leave the star, although with somewhat less energy.

But if the core of the collapsing star is squeezed into a very small volume, the force of gravity on its surface is very great. Suppose the core is squeezed down to a radius of a few miles. At that point, the mass is so compact that the force of gravity at the surface is billions of times stronger than the force of gravity at the surface of the sun. The tug of that enormous force prevents the rays of light from leaving the surface of the star; like the ball thrown upward from the earth, they are pulled back and cannot escape to space. All the light within the star is now trapped by gravity. From this moment on, the star is invisible. It is a *black hole in space*.

Inside the black hole, the contraction continues, piling up matter at the center in a tiny, incredibly dense lump. According to current knowledge in theoretical physics, this is the end of the star's life. The star's volume becomes smaller and smaller; from a globe with a two-mile radius it shrinks to the size of a pinhead, then to the size of a microbe, and, still shrinking, passes into the realm of distances smaller than any ever probed by man. At all times the star's mass of a thousand trillion trillion tons remains packed into the shrinking volume. But intuition tells us that such an object cannot exist. At some point the collapse must be halted. Yet, according to the laws of twentieth-century physics, no force, no matter how powerful, can stop the collapse. The implication is that the laws of physics must be modified at extremely short distances in a manner that prevents particles from coming infinitely close together. Here is a hint of the impending discovery of a new force or a new law in nature, one that may transform the world of the future as the discovery of the nuclear force transformed the world of the twentieth century. The study of the stars may yet dominate the affairs of men.

Black holes have many other remarkable properties. For example, the force of gravity within a black hole not only prevents light from escaping; it also prevents all physical objects from getting out of the hole. This property of black holes is another prediction of Einstein's theory, which asserts that no object can travel faster than light. If the black hole's gravity is so powerful that light cannot break its grip and escape to space, material objects cannot escape either. Everything inside the black hole is trapped there forever.

Any ray of light or physical object that enters the black hole from the outside is also trapped; it can never emerge again. The interior of the black hole is completely isolated from the outside world; it can take in objects and radiation, but cannot send anything back. It is almost as though the material inside the black hole no longer belongs to our universe. In fact, some astronomers have suggested that black holes are places from which matter is pouring out of our universe and into some other universe that exists in another dimension.*

Since black holes capture any material they encounter, the size of a black hole will always tend to increase in time. The black hole is, in a sense, insatiable. As more matter enters it, the strength of its gravity increases, and therefore its sphere of influence expands. This property does not imply that black holes act as gravitational vacuum cleaners, pulling in all the matter around them. A spaceship can pass by a black hole safely as long as it does not come very close. However, if a ship heads directly for a black hole, it will enter the black hole and vanish.

What would happen to an astronaut who fell into a black hole? The properties of black holes seem to suggest that he would be crushed by gravity. In actual fact he would be torn apart, because the part of his body closest to the center of the black hole would be pulled by a stronger gravitational force than any other part. Suppose, for example, the astronaut entered feet first; then his feet would be pulled more strongly than his head, and feet and head would tend to separate. The astronaut would feel as though he were stretched on a rack; a few thousandths of a second after entering the black hole, he would be dismembered; after a few more thousandths of a second, the individual atoms of his body would be broken into their separate neutrons, protons, and electrons; finally the elementary particles themselves would be torn into fragments whose nature is not yet known to the physicists.

The black hole is the most bizarre and unbelievable object that has ever been conceived by the scientific mind. Yet the theory of rela-

---

*The tiny region from which matter issues forth into the other universe is called a white hole in this theory.

tivity, combined with calculations on the evolution of massive stars, assures us that black holes must exist; whenever a very massive star undergoes a supernova explosion, a black hole must be left behind. How can this prediction be tested?

A test would seem to be impossible, since the black hole by its nature cannot be seen. However, recent discoveries made with satellites provide tentative evidence that black holes actually do exist. The satellites are equipped to measure x-rays coming to the earth from space. These x-rays cannot be seen on the ground because the earth's atmosphere screens them out. One very powerful source of x-rays discovered by the satellites seems to be coming from a bright, conspicuous star. However, a very careful observation shows that the x-rays are not coming from the star itself, but from an invisible object quite close to it.

The invisible object is thought by astronomers to be a black hole circling around a normal star. According to their calculations, the powerful gravitational pull of the black hole tears streamers of gaseous matter away from the neighboring star. The atoms of gas speed toward the boundary of the black hole, colliding violently on the way, and the collisions produce an intense stream of x-rays. Astronomers have tried to invent other theories to explain the mysterious x-rays; however, nothing fits all the known facts as well as the black hole.

With the tentative evidence for the existence of black holes, the final pages in the life story of the stars appear to have been written. But the story has an epilogue. When a supernova explosion occurs and the outer layers of the star are sprayed out into space, they mingle with fresh hydrogen to form a gaseous mixture containing all 92 elements. Later in the history of the galaxy, other stars are formed out of clouds of hydrogen which have been enriched by the products of these explosions. The sun is one of these stars; it contains the debris of countless supernova explosions dating back to the earliest years of the Galaxy. The planets also contain the debris; and the earth, in particular, is composed almost entirely of it. We owe our corporeal existence to events that took place billions of years ago, in stars that lived and died long before the solar system came into being.

THE BIRTH
AND DEATH
OF STARS

A Cloud Pulled Together By Gravity

The Manufacture of Elements Within Stars

The Birth of a Star in Space

The Structure of a Star Late in Life

Dying Stars

A CLOUD PULLED TOGETHER BY GRAVITY. It is believed that stars are born in the swirling clouds of hydrogen gas which fill all of space. If the atoms in one region of such a cloud come together by accident or are forced together by the pressure of the surrounding clouds, the force of gravity pulls the atoms still closer together, forming a condensed pocket of gas *(below, left)*. The continuing action of gravity compresses the pocket of gas *(below right)*. As a result of the compression, the temperature at the center rises; after about 10 million years, when the temperature has reached the critical value of 20 million degrees, nuclear reactions set in, in which vast amounts of energy are released. The onset of these reactions marks the birth of a star. The release of nuclear energy halts the further collapse of the star. The energy passes to the surface and radiates into space in the form of light and heat.

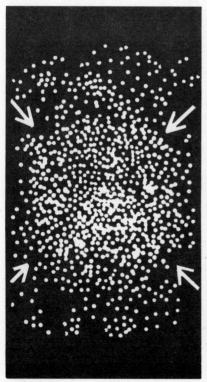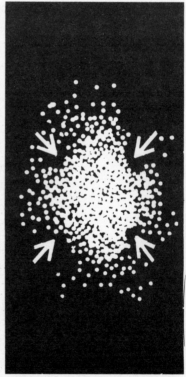

THE MANUFACTURE OF ELEMENTS WITHIN STARS. Most of a star's energy is derived from nuclear reactions in which hydrogen nuclei, or protons, fuse to form helium nuclei. Four protons unite to form one helium nucleus in this reaction, shedding, at the same time, two units of positive electric charge *(below)*.

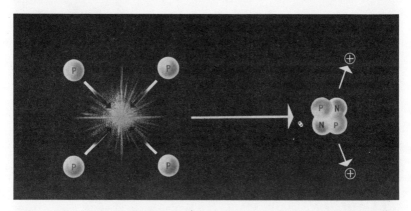

The fusion of hydrogen to form helium continues for 99 percent of a star's lifetime. When the hydrogen is substantially depleted the star again collapses until its center reaches a temperature of 200 million degrees Fahrenheit. At this temperature helium nuclei fuse to form the nucleus of carbon *(below)*. Oxygen and still heavier elements are formed after carbon. In this way, the elements of the universe are synthesized, step by step, out of the basic building block of hydrogen.

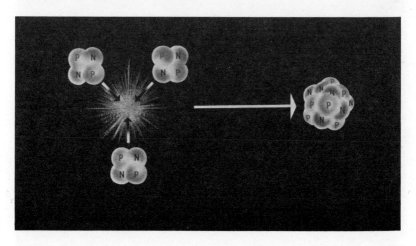

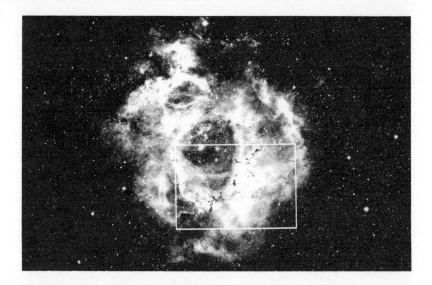

THE BIRTH OF A STAR IN SPACE. The luminous cloud of gas *(above)*, contains many small, dark areas, which can be seen clearly in the enlarged section *(below)*. These dark regions are dense globules of gas in which new stars are about to form. Arrows point to two of the star-forming regions. The globules are dark because they contain substantial amounts of dust particles, which absorb the light passing through them.

THE STRUCTURE OF A STAR LATE IN LIFE. At the beginning of a star's life it burns hydrogen to make helium. In the later stages of its life, helium burns to make carbon. The illustration *(below)* shows the interior of the star during the later period in which carbon is accumulating at its center. The carbon is surrounded by burning helium; the helium is surrounded, in turn, by a layer of hydrogen that has been burning since the star was born.

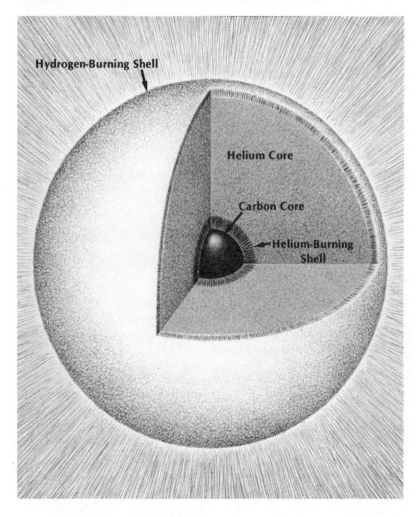

A DYING STAR. Small stars collapse and fade away at the end of their lives, when their nuclear energy is spent. Massive stars, on the other hand, explode at the end of their lives. These exploding stars, called supernovas, flare up with a brilliance billions of times greater then the sun. If a supernova is located nearby in our galaxy, it appears suddenly as a very bright star, visible in the daytime sky.

One of the earliest reported supernovas was observed by Chinese astronomers in A.D.1054. Today at the position of this supernova there is a large cloud of gas known as the Crab Nebula, shown in the photograph (below), which is expanding outward at a speed of 1000 miles per second.

The supernova explosion sprays the material of the star out into space, where it mingles with fresh hydrogen to form a mixture containing all 92 elements. Later in the history of the galaxy, other stars are formed out of clouds of hydrogen which have been enriched by the products of these explosions. The sun, the earth, and the beings on its surface — all were formed out of such clouds, containing the debris of supernova explosions dating back billions of years to the beginning of the Galaxy.

The Veil Nebula (opposite) is the gaseous remnant of a star 2500 light years away, which exploded into a supernova about 50,000 years ago.

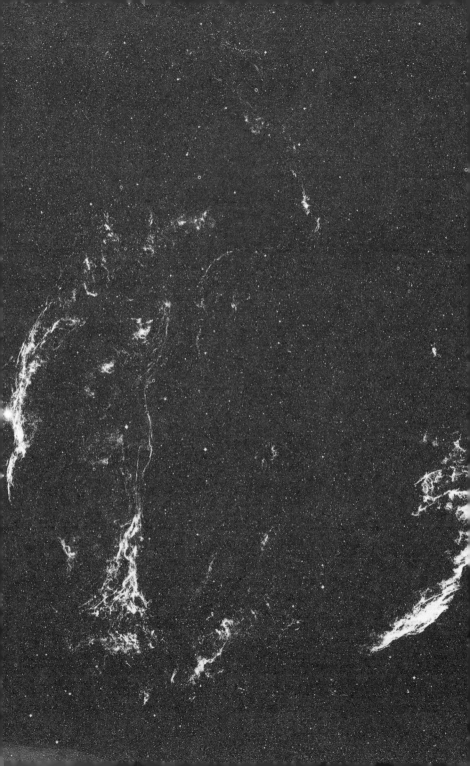

steady state theory
Big Bang theory
Oscillating   "

6   THE BEGINNING AND THE END

A star is born out of the condensation of clouds of gaseous hydrogen in outer space. As gravity pulls together the atoms of the cloud, its temperature rises until the hydrogen nuclei within it begin to fuse and burn in a series of reactions, forming helium first, and then all the remaining substances of the universe. The elements of which our bodies are composed were manufactured in this way, in the interiors of stars now deceased, and distributed to space when these stars exploded. Subsequently, these elements were drawn together again in the cloud of gas out of which the sun and the earth condensed. If the sun explodes at the end of its life the planets will be consumed, and their substance once again distributed into space, to be reincarnated in another solar system as yet unborn.

This beautiful theory allows the universe to go on forever in a timeless cycle of death and rebirth, but for one disturbing fact. Fresh hydrogen is the essential ingredient in the plan, but with the passage of time the supply of fresh hydrogen dwindles; as the old stars go out, one by one, fewer and fewer new stars are formed to replace them. Stars are the source of the energy by which all beings live. When the light of the last star is extinguished, life must end throughout the universe.

Other evidence suggests that the universe is changing in an irreversible way. All the galaxies seem to be moving away from one another at very high speeds. Those most distant from us are receding at the extraordinary speed of 150,000 miles per second, which is close to the velocity of light. The universe appears to be blowing up before our eyes, as if we were witnessing the aftermath of a gigantic explosion. As the galaxies fly apart and the distances between them increase, space grows emptier, and the density of matter dwindles to nothing. It seems, that, albeit slowly, the universe is approaching an end.

Some years ago, Thomas Gold, then a graduate student at Cambridge University, made a proposal which voids these morbid predictions. He suggested that fresh hydrogen is steadily created throughout the universe *out of nothing*. The freshly created hydrogen would provide the ingredients for the formation of new stars to replace the old. At the same time it would fill up the spaces left by the movement of the galaxies away from one another. Thus, the creation of matter out of nothing, as proposed by Gold, could restore the universe to a state of perpetual balance, without beginning and without end.

Gold mentioned his idea to Herman Bondi and Fred Hoyle, two English astronomers, who joined him in working out the consequences. They asked themselves, how much hydrogen should be created per year in order to keep the density of matter constant everywhere, as the universe expands? According to their calculations, the expanding universe remains in a steady state with a constant density of matter, if one hydrogen atom is created per year in a volume equal to that of the Empire State Building.

This is a very modest rate of creation, but it violates a cherished concept in science—the principle of the conservation of matter—which states that matter can be neither created nor destroyed. It seems difficult to accept a theory that ignores such a firmly established fact of terrestrial experience. Yet, the proposal for the creation of matter out of nothing possesses a strong appeal, for it permits us to contemplate a universe that extends into the past and the future without limit, a universe that renews itself *in perpetuum*.

Another cosmology, which offers strong competition to the steady-state theory, makes no attempt to dodge the implications of the expanding universe. This cosmology, appropriately named the big-bang theory, proposes that the expansion is in fact the consequence of an actual explosion which took place a long time ago. Father Lemaître, a Belgian astronomer educated as a Jesuit priest, and

George Gamow, a Russian-born physicist who emigrated to the United States in 1936, are the scientists most prominently associated with this theory. In 1931 Lemaître proposed that the universe began its existence as a condensed droplet of matter at an extremely high density and temperature. Later Gamow named this primordial egg "ylem"—the name that Aristotle gave to the basic substance out of which the ancients believed all matter was derived. Internal pressures within this hot, dense droplet, containing all of the matter and radiation of the universe, caused it to expand rapidly. As it expanded, its temperature and pressure dropped. In the first few minutes of its existence the temperature was many millions of degrees, and all the matter within the droplet consisted of the basic particles—electrons, neutrons and protons. Any combination of these particles to form nuclei or atoms would quickly be disintegrated under the smashing impact of the violent collisions which occur at such high temperatures. However, as the universe continued its expansion, and the temperature of ylem dropped further, the protons and neutrons began to fuse together to form hot nuclei of the element helium.

With the further passage of time, the matter of the universe cooled and condensed into galaxies, and within the galaxies, into stars. After some billions of years of continued expansion, the universe reached the state in which it exists today. Knowing how far apart the galaxies now are, and how rapidly they are moving away from one another, we can calculate backward in time to the moment at which the expansion began. In this way, astronomers arrive at the conclusion that the universe began its existence 20 billion years ago. At the present time, in the words of Lemaître,

> "the evolution of the world can be compared to a display of fireworks that has just ended: some few red wisps, ashes and smoke. Standing on a cooled cinder, we see the slow fading of the suns, and we try to recall the vanished brilliance of the origin of the worlds."

Is the big-bang cosmology correct? Can everything in our universe be traced back to one explosive moment? Some years ago, Ralph Alpher and Robert Herman, two nuclear physicists, pointed out an

aspect of the big-bang theory which could be tested experimentally.*
According to the theory, the universe came into existence as a
droplet of hot, dense matter. This early universe must have been a
fireball filled with an intense, brilliant radiation. As the universe expanded, the intensity of the fireball diminished. The calculations of
Alpher and Herman indicated that if the big bang really did occur, a
remnant of the fireball should exist today, and should be detectable
with a sensitive radio antenna.

For some reason, no one took the suggestion of Alpher and Herman seriously enough to look for the fireball radiation. It is puzzling
that this opportunity to discover how the world began failed to
engage the attention of astronomers at the time. Perhaps the reason
is that the project was too bold in its concept to seem scientifically
sound.

In any event, twenty years elapsed and then, at the Bell
Laboratory in New Jersey, two physicists, Arno Penzias and Robert
Wilson, stumbled across the fireball radiation quite by accident.
Penzias and Wilson were not looking for proof of the beginning of
the world; they were measuring the intensity of radio waves in a
large antenna which had been set up for a completely different pur-
pose. Their measurements revealed a puzzling radiation which came
to their antenna from all parts of the sky. Penzias and Wilson were
unable to explain the source of this radiation until a friend sent them
an article that described the theoretical properties of the cosmic
fireball. The rest is scientific history.

Subsequently, other physicists and astronomers confirmed the existence of the fireball radiation. Their confirmation goes beyond the
mere presence of the radiation; the latest measurements show that
the wavelengths, or colors, present in this radiation exactly match
the pattern of wavelengths expected for the flash of light and heat
emitted in a great explosion. This pattern of wavelengths is the
fingerprint of the big bang, blurred after 20 billion years but

*Alpher and Herman drew their inspiration from the work of George Gamow on
the primordial egg.

unmistakable.

The measurement of the wavelengths in the cosmic radiation convinced nearly all astronomers that the big bang really did occur. No competing explanation for the radiation was provided by the steady-state cosmologists. As a consequence, their theory fell into disrepute.

The scientific community gave its final approval to the big-bang cosmology in 1978, when Penzias and Wilson received the Nobel Prize for discovering a relic of the beginning of the universe. But the concept of the big bang creates very serious problems for science. If this theory is correct, the universe began suddenly, in a flash of light and heat, some 20 billion years ago. What are we to make of such a picture? If there was a beginning, what came before? When all the stars burn out, what comes after? Some cosmologists say that gravity, acting on all the outward moving galaxies, will halt their expansion and draw the elements of the universe together, compressing them into a dense, hot mass, until another big bang sends them hurtling outward; and again they fall inward. . . .

In this way, oscillating between expansion and contraction, the universe is preserved throughout eternity, and the difficult questions of beginning and end are avoided. However, there is evidence against the oscillating theory. The force of gravity in the universe has turned out to be at least ten times too small to halt the outward motion of the galaxies, According to the available facts, the universe will expand forever.

There the matter rests for the moment. Scientists have exposed very interesting features of the Great Plan—the birth of stars, the assemblage of the elements within the stars out of the three basic particles, and their dispersal to space in supernova explosions; but science offers no satisfactory answer to one of the most profound questions to occupy the mind of man—the question of beginning and end.

THE NEW
COSMOLOGY

The Expanding Universe

The Big-Bang Cosmology

Steady-State Cosmologists

The Primordial Fireball

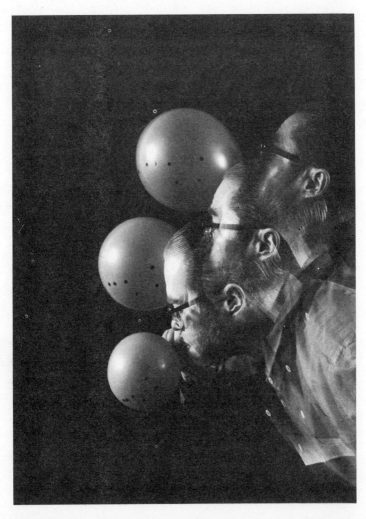

THE EXPANDING UNIVERSE. Astronomical observations indicate that all the galaxies in the universe are moving away from one another at very high speeds, just as if they were on the surface of a rapidly expanding balloon *(above)*. The nature of the expansion is difficult to conceive when it occurs in the three dimensions of the real world rather than on the two-dimensional surface of a balloon, but the principle is the same.

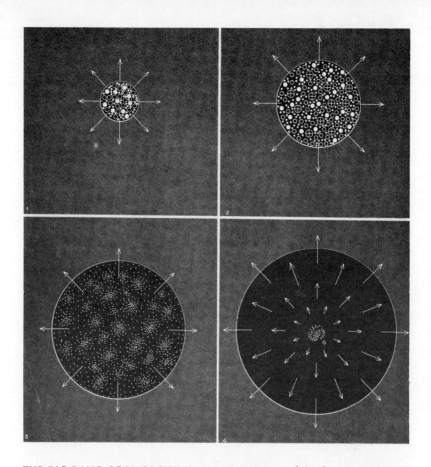

THE BIG-BANG COSMOLOGY. In an attempt to explain the expansion of the universe, theorists have proposed that 20 billion years ago the universe was a hot, dense cloud of matter and energy. The highly compressed universe expanded outward with explosive violence. The first figure shows the universe one second after the initial explosion that marked its birth. At this time the temperature of the universe was 10 billion degrees. In the second figure, three minutes later, the universe is still expanding, but the temperature has dropped to one billion degrees. Neutrons and protons are fusing together to form helium nuclei. In the third figure, a billion years later, the gases of the universe are considerably cooler, and atoms have formed and have started to condense into galaxies, and into stars and planets within each galaxy. The last figure shows the expanding universe as it would appear from any one of these galaxies. Each galaxy seems to be at the center of the expansion because it sees all neighboring galaxies moving away from it.

STEADY-STATE COSMOLOGISTS. Another effort to explain the expansion of the universe was made by the English cosmologists, Herman Bondi, Tom Gold and Fred Hoyle. Their theory, called the steady-state cosmology, attempted to avoid the difficulties stemming from a theory in which the universe has a beginning at a definite point of time, by proposing that hydrogen is continuously created out of nothing throughout the universe. They suggested that the newly created hydrogen would fill the void left by

the expansion of the universe and, at the same time, provide fresh material for the formation of stars. Thus the universe renews itself indefinitely, without beginning and without end.

In the photograph Hoyle *(left)* and Gold *(middle)* are shown engaged in conversation with Freeman Dyson of the Institute for Advanced Study in Princeton, at a conference held at the Goddard Institute for Space Studies in New York.

THE PRIMORDIAL FIREBALL. In 1965 strong evidence appeared against the steady-state theory and in support of the big-bang cosmology. Around that time, Robert Dicke, a Princeton physicist, started to look for the radiation left over from the birth of the universe. He was the first scientist to take up the suggestion made by Alpher, Herman and Gamow, nearly twenty years before, that this radiation might still be present and could provide a proof of the big-bang theory.

As Dicke began to construct an apparatus to search for the fireball radiation, two radio astronomers at the Bell Laboratories, Arno Penzias and Robert Wilson, stumbled across it while measuring radio signals from space, using a large antenna set up by A.T.&T. for the communications satellite program.

Robert Dicke

Arno Penzias *(right)* and Robert Wilson in front of the A.T.&T. antenna.

7 THE ORIGIN OF THE SOLAR SYSTEM

The origin of the solar system is less of a mystery than the origin of the universe, but it is still not a clearly understood event. When I was in high school I was taught the theory that the planets came into being as by-products of a catastrophic event in which the sun collided with a passing star. The force of gravity tore huge streamers of flaming gas out of the bodies of the two stars during this encounter. As the intruding star receded into the distance, some of these streamers of gaseous material were attracted by the sun's gravity and captured into orbits circling around it. The earth condensed out of one of these streams of hot gas to form a molten mass, on whose surface a crust formed and gradually hardened with the passage of time.

It is easy to calculate the probability that the solar system originated in this way. The likelihood of a collision between the sun and another star depends on the size of the sun, and on the distance between it and its neighbors, Stars, large though they are, are very small in comparison with the average distances that separate them. The sun, for example, is one million miles in diameter, and 25 trillion miles from its nearest neighbor. For that reason, the chance of a collision between two stars is very small; a calculation shows, in fact, that only a few collisions could have occurred during the history of the Galaxy. Indeed, the planets in our solar system may be the only ones in existence, according to the collision theory.

A very different prediction comes out of the modern theory of the formation of the stars. This theory asserts that planets are formed as a natural accompaniment to the birth of a star. As the gas cloud of the star-to-be contracts under the inward force of its own gravity, the density of the gas increases, and atoms and molecules collide and occasionally stick together. Gradually, small fragments of solid matter accumulate in the cloud as a result of these collisions. What happens next is not well understood. According to one theory of the birth of the planets, when the fragments of solid matter are still quite small they are swept along with the surrounding gas in its motion around the center of the cloud. However, when the fragments become large enough—roughly basketball-sized or larger—they are too massive to be carried with the gas. They begin to "fall" out of the gas cloud toward the central plain of the cloud of matter around the newly forming star, where the density of the gas is greatest and the pull of gravity is strongest. In the course of time, the basketball-sized objects become concentrated in a thin disk of matter rotating around the young star at the center.

The situation is analogous to that of a cloud of moist air in which droplets are condensing; when the water droplets are small, they move with the air currents, but when they have grown to a sufficiently large size, they fall out of the cloud in the form of rain.

According to this picture, when our solar system was young it consisted of a relatively dense, hot cloud at the center—which eventually formed the sun—surrounded by a large number of solid bodies and planetesimals of various sizes. With the further passage of time, the central cloud continued to contract and grow hotter, until finally its temperature reached the critical level of 20 million degrees needed for the ignition of thermonuclear reactions. This point in the contraction marked the true birth of the sun. That occurred about 4.6 billion years ago.

For millions of years, throughout the period in which the sun was forming, the aggregation of small bodies into larger ones continued until the planets were completely formed. The last stage in the formation of the planets may have occurred somewhat before the sun was born, or later. We are not sure which came first, sun or planets.

The 34 moons in the solar system could have been formed in the same manner, as still smaller condensations around their parent planets.

The condensation theory of the origin of the planets implies that planets are formed nearly every time a star is formed. They must, therefore, be very common objects in the universe; in fact, there must be billions of planets in our galaxy alone. Much of the present discussion regarding the origin of the planets rests on this single prediction. Unfortunately, the prediction cannot be tested directly by astronomers, because a planet circling a star in another solar system is too faint an object to be seen in the largest telescopes on the earth. If a 200-inch telescope were placed in an orbiting satellite or in an observatory on the moon, far from the obscuring effects of the earth's atmosphere, it would be just barely able to detect a Jupiter-sized planet circling one of the stars near the sun. Telescopes of this size will not be placed in orbit, or on the moon, for some years to come.

According to the condensation theory, stars and planets are both formed by the contraction of a cloud of gas under gravity. In that case, why are stars so different from planets? The answer is connected with their masses. As the star-cloud, or the planet-cloud, contracts, its temperature rises. If the cloud is very massive the temperature reaches the value of 20 million degrees, at which nuclear burning commences, and a star is born. If the cloud is small, the temperature fails to reach the critical level, and the condensed cloud remains an inert body without internal sources of nuclear energy; that is, it becomes a planet.

The smallest cloud of gas that will create a star is approximately one-twentieth the mass of our sun. Jupiter is a very large planet, 318 times more massive than the earth, but still it falls short, by a factor of about 50, of having the mass needed to form a second star in our solar system.

It is possible that some of the clouds of gas which condense in outer space are too small to produce the temperatures needed for nuclear burning and, therefore, cannot become stars; yet they cannot properly be called planets, because a planet is a body that is bound to a star by gravity and circles in orbit around it. Such relatively cold, planet-sized bodies, free of the gravitational influence of any star, may exist around us in considerable numbers. Since they do not shine by their own light, and are too far away from any star to be visible to us by its reflected light, we have no way of detecting these "free planets" at present. We may stumble across one of them when our unmanned spacecraft begin to explore the regions lying beyond the limits of the solar system.

These thoughts are derived from the condensation theory of the origin of the solar system, for which no definite proof has been supplied. Nonetheless, the theory is generally accepted because it fits so naturally into the latest ideas on the birth of stars.

Yet, in spite of its successes, the condensation theory has failed to give us a clear picture of the events that created our own planet. We know less about the processes that formed the earth than we do about the birth of distant stars. The skies contain old stars, young stars, and some that are just in the process of being born—all directly available for examination in our telescopes. Through them we have learned the story of the formation of the stars and their evolution into the red giants and the white dwarfs. But the early years of the earth's history are shrouded in mystery, because volcanic eruptions and erosion by wind and running water have combined to erase the record of the earth's past. The exploration of the moon has yielded important new information regarding the early history of the planets, but the moon is the same age as the earth and we can never learn what events actually brought it into being; we can never learn as much as we could if we were to be present at the very moment of a planet's birth.

Perhaps we will discover a planet younger than the earth one day, if ever we are able to leave this solar system and voyage to other stars, but such voyages are not in prospect in the foreseeable future. For the present we are doomed to ignorance of the conditions which existed on the earth during its formation and its early years. We do not know the temperature at the surface of the young earth; the gases which floated in its atmosphere; and the chemicals which were dissolved in the primitive oceans. These facts are unknown, and perhaps unknowable, and yet they are of great interest, because they are bound up with the question of life's origin. The earliest traces of living organisms discovered thus far—residues of bacteria and simple plants—are found in rocks about 3 billion years old. When these organisms were alive the earth had already existed for more than one billion years. During that period of a billion years life developed here. *What were the conditions under which it arose?*

We believe that at the beginning there was only a cloud of gaseous hydrogen, mixed with small amounts of other substances. Out of this cloud grew the sun, the planets, and the creatures which walk on the surface of the earth. It was the parent cloud of us all. At its center existed a dense, hot nucleus which later formed the sun. The outer regions—cooler and less dense—gave birth to the planets.

Out of what materials were the planets formed? The bulk of the parent cloud must have been composed of the light gases, hydrogen and helium, because they are the most abundant elements in the universe. Other elements relatively abundant in the universe, although less so than hydrogen and helium, are carbon, nitrogen and oxygen, metals such as iron, magnesium and aluminum, and silicon. These substances must also have been present in relatively great abundance in the parent cloud of the planets. No doubt the remaining eighty-odd elements were also represented, but in smaller amounts.

All the familiar chemical compounds of these substances would have formed in the cloud in a relatively short period of time. Hydrogen combines readily with oxygen to form molecules of water vapor; hydrogen also combines with nitrogen to form molecules of ammonia gas, and it combines with carbon to form methane, also called marsh gas, which is used extensively today for cooking. Carbon and oxygen combine to form carbon dioxide. Considerable amounts of each of these compounds must have formed in the parent cloud. However, they were probably not present in the form of gases, because of the low temperature—about 100 degrees Fahrenheit below zero—prevailing in the region of the cloud out of which the earth was formed. At this temperature they congealed into a slushy mixture of water, ammonia and methane ice in liquid and solid form, plus solid carbon dioxide—dry ice. The other elements that were present in abundance—silicon, aluminum, magnesium and iron—combined with oxygen to form grains of rocklike materials and metallic oxides.

92

Solid carbon dioxide = dry ice

These, then, are the substances out of which the planets condensed: a Neapolitan sherbet of frozen water, ammonia and methane, plus various kinds of rocky substances—all immersed in a gaseous cloud of hydrogen and helium.

When the planets first condensed out of this mixture of gases and solid matter, the bulk of their mass should have consisted of hydrogen and helium. The giant planets—Jupiter, Saturn, Uranus and Neptune—are in fact composed mostly of these light gases, but for some reason the earth and its nearest planetary neighbors lack them. Why they are scarcer on the earth than on Jupiter is a mystery. Some students of the subject say that they were blasted away by rays of the sun, which was much more brilliant than it is today. Others say the opposite: that all the light gases in the inner regions were drawn into the body of the primitive sun as it contracted, and the earth was formed out of the rocky materials left behind.

Whatever may have caused the departure of hydrogen and helium, it is clear from the scarcity of these gases on the earth, that they had disappeared from the neighborhood of the earth's orbit to a large extent by the time our planet had started to condense. There remained only particles of rock and small amounts of ice, which circled in orbit around the sun, each a miniature planet in its own right.

Occasionally collisions occurred between neighboring particles in the course of their circling motion. Some collisions were gentle, and the particles stuck together. In this way, in the course of millions of years, small grains of rock gradually grew into larger ones. Some pieces of rock became large enough to exert a gravitational attraction on their neighbors. These were the nuclei of the modern planets. Once they had grown large enough to attract other particles by their own gravity, they quickly swept up all the materials in the space around them, and developed into full-sized planets in a short time.

The complete process of planet formation went on over a period of perhaps 50 million years, proceeding with extreme slowness at first, and then with rapidly increasing momentum in the final stages. At the end, all the matter of the solar system was gathered into the existing planets, and only a few atoms of gas remained in the space between. This is the situation in the solar system as it exists today.

THE FORMATION OF THE SUN AND PLANETS *(opposite)*. The solar system is believed to have begun its life as a diffuse cloud of gas held together by the force of its own gravity (1). The original radius of the cloud was approximately 10 trillion miles.

With the passage of time the attraction of gravity drew the atoms of gas together and the cloud contracted (2).

Over a period of time, which is difficult to determine but was probably between 10 and 20 million years, the dense center of the cloud gradually shrank to approximately the size of the present sun (3). At this point the temperature within the central region was high enough to ignite nuclear reactions, in which hydrogen fused to form helium, releasing energy and marking the birth of the sun.

In the cooler and less dense regions surrounding the primitive sun, smaller condensations developed under the force of gravity, giving rise to the planets (4).

1

10 trillion miles

2

10 to
20 million
years later

3

4

MARS VENUS

MERCURY SUN

EARTH

8 THE MOON: ROSETTA STONE OF THE PLANETS

We estimate that the earth and the moon, along with the rest of the solar system, were formed 4.6 billion years ago. Sometime in the first billion years, life appeared on the earth's surface. Slowly, the fossil record indicates, living organisms climbed the ladder from simple to more advanced forms until—perhaps a million years ago—the threshold of intelligence was crossed.

Earthbound creatures can never learn how that happened, or what conditions led to the emergence of life, because the record of the earth's early years has been wiped out. The air and water that make our planet livable have worn down the oldest rocks and washed away their remains into the oceans, while mountain-building activity and volcanic eruptions have churned the surface and flooded it repeatedly with fresh lava, removing the remaining evidence. These natural forces have entirely removed the primitive materials that lay on the earth's surface when it was first formed. No rocks have ever been found on the earth that are older than 3.7 billion years. We know nothing of what happened on our planet from the time of its formation, 4.6 billion years ago, to the time when these oldest rocks were laid down. The critical first billion years, during which life began, are blank pages in the earth's history.

But on the moon there are no oceans and atmosphere to destroy the surface and there is relatively little of the mountain-building activity that rapidly changes the face of the earth. Over large areas, the materials of the moon's surface are as well preserved as if they had been in cold storage. The moon offers the best chance of recapturing the lost record of the earth's past.

A casual inspection of photographs of the moon immediately confirms that this small planet has retained the record of early events in its history with exceptional fidelity. The photographs show countless craters, most of which have been produced by the impact of meteorites raining down on the moon for billions of years. Many craters are circled by ramparts ranging up to 10,000 feet in height. Some of these ramparts must be a billion years old or more, yet photographs taken with a telescope clearly indicate that they have been preserved almost unchanged, with little of the original material worn away. Meteorites have collided with the earth throughout its history, just as they have collided with the moon, and they have produced similar craters; but all traces of the older craters are gone because on the earth various agents of erosion, of which the most important is running water, move materials from one place to another, leveling the crater walls and filling in the pits and hollows. Only the scars of the most recent collisions, such as the Arizona meteorite crater, formed about 30,000 years ago, are still visible on the earth.

Pictures of the moon taken by early NASA spacecraft provided proof that the extent of erosion on the moon has indeed been very small. The photographs showed lunar features as small as a few feet in diameter. The clarity of the spacecraft photographs produced jubilation among astronomers, who had been straining to see through the earth's atmosphere like drivers peering at the road through a rain-spattered windshield. Many craters which had never been seen before in photographs taken with telescopes on the earth were visible, ranging from a few feet in diameter up to hundreds of feet. These small craters must have existed on the earth as well, but were wiped out almost immediately by the wearing effect of winds and running water. On the moon, such craters can last for a billion years, and even the shallow footprints of the Apollo astronauts, no more than six inches deep, will still be visible ten million years from now.

The moon's resistance to the ravages of time endows its barren landscape with a unique value. The rocks that litter the moon's surface probably contain no life; we know they contain very little gold or silver; nonetheless, they are scientifically priceless because of the revelations they can offer regarding the early years of the solar system.

The expectation that the moon will supply these revelations is based on the "cold-moon" theory, first proposed by Harold Urey. This theory states that the moon is relatively cold, and very different in that respect from the earth. The earth is hot inside, with many small pockets of molten rock scattered throughout its interior. When one of these pockets is connected to the surface of the earth by a crack or a duct in the solid crust, volcanic upheavals occur, accompanied by great floods of hot, liquid rock. Throughout the earth's history, its surface has been made over by these convulsions in the interior, which continually bring up new material that spreads out over the surface as lava and congeals to form fresh rock. The older rocks are buried by the congealed lava. As a consequence, most of the rocks on the earth's surface are relatively young—not more than a few hundred million years old.

The moon's surface, on the other hand, has not been flooded with lava for billions of years, according to the cold-moon theory. Its old rocks have not been buried, and they are lying on the surface, easily accessible to the geologist.

Prior to the first lunar landing, many geologists held a very different opinion about the moon. They saw it as a hot planet similar to the earth, and argued that its surface, like that of the earth, had been covered frequently with floods of molten lava, which buried the record of its past. According to these "hot-moon" scientists, the moon had little to contribute to our knowledge of the early years of the solar system.

Attitudes hardened as the day of the lunar landing approached, but partisans of both viewpoints believed that Apollo 11 would settle the controversy. A feeling of intense excitement gripped us as we awaited the return of the priceless rocks. Finally, the day came on which the first box of rocks was scheduled to be opened. Tension mounted in the viewing room of the Lunar Receiving Laboratory as scientists and reporters watched the Laboratory staff conduct an elaborate ritual devised by suspicious earthlings to protect themselves and their environment from their first contact with an alien planet. There was only a minute probability that the lunar rocks would contain a lethal microorganism against which life would be defenseless, but the potential danger was great; hence painstaking precautions were taken. Hot-moon and cold-moon scientists were present. We knew that weeks or months would pass before the experiments that could determine the ages of the rocks would be completed, but we also knew that the moment of truth was close at hand.

The results of the age measurements were revealed at a unique conference of lunar scientists in Houston in 1970. One hundred and forty-two Apollo 11 research teams presented their reports to more than a thousand colleagues and reporters who had come to Houston from all parts of the world. Working with small amounts of the precious material, averaging no more than a thimbleful, they had subjected the rocks to every possible variety of investigations, ranging from absurdly simple operations—equivalent to kicking a tire—to complex, delicate laboratory analyses; they had squeezed the rocks under high pressure, heated them to the melting point, examined them under microscopes with polarized light, bombarded them with x-rays, and finally taken them apart, atom by atom.

The conference sessions were dreary, as oversized scientific gatherings usually are. Hundreds of people sat in the darkened hall, numbed by exposure to a rapid succession of graphs and charts. The volume of facts and figures paralyzed the mind, yet an undercurrent of excitement gripped us, for we were witnesses to an historic occasion—the first discussion by earthlings of alien materials brought home from another planet.

The highlight of the conference came during the opening session: All the rocks turned out be billions of years old, and some were 4.6 billion years old—*as old as the solar system.*

The significance of that single result cannot be exaggerated. It meant that some of the dusty fragments had lain on the moon since the earliest years of the solar system, and that through them we could transport ourselves backward in time to the moment when the planets were freshly condensed out of the swirling clouds of the solar nebula. What elements went into the newly created earth? How hot was its surface? How dense was its primitive atmosphere? Were molecules present, of the kind that could lead to the development of life? Not a single fragment of the primitive earth can be found today to answer these questions; but pieces of very old moon are available; in fact, judging by the Apollo samples, they are abundant.

There was rejoicing in the camp of the cold-moon scientists. They had been prepared to make a long, patient search for those rare objects, the oldest rocks in the solar system, and now they found these precious antiquities strewn about on the moon, waiting to be picked up in the very first hit-and-miss collection. They felt like prospectors who, hoping to find a few bits of gold dust, had stumbled onto a field carpeted with nuggets.

And there was more. The lunar highlands—a rough, mountainous no man's land of jumbled rocks and meteorite craters—turned out to be the original, pristine surface of the moon, dating back to the beginning of the solar system. The lunar highlands provide a clue to the way the earth must have looked at the beginning of its life, before the oceans and continents appeared.

The rocks from the lunar maria, or seas, on the other hand, were somewhat younger. The lunar maria are the large, relatively dark areas that make up the features of the Man in the Moon.* According to the measurements of the ages of the rocks from the maria, these rocks were laid down when the moon was already a billion years old.

A detailed study of the rocks from the maria revealed still more about the history of the moon and, by implication, about the history of the earth. All the samples collected from the lunar maria turned out to be a kind of dark rock known as basalt. Basalt is a congealed lava, formed when molten rock rises to the surface of a planet and

---

*When Galileo, the first man to look at the moon through a telescope, turned his primitive instrument on that body in 1609 he saw dark, smooth areas resembling oceans, which he called *maria*. This Latin world for seas has remained, but we know now that the resemblance to bodies of water is illusory. The lunar seas contain no water; no storms rage across the plains; no streams flow down from the highlands. The moon is a dry planet.

solidifies. The presence of basalt means the moon was once volcanically active, with pockets of molten rock in its interior and floods of lava pouring out at the surface. These floods of lava must have formed the lunar maria.

According to the ages of rocks from the maria, eruptions repeatedly occurred over the course of several hundred million years; then, about three billion years ago, volcanic activity ceased. Since that time, no new floods of lava have appeared on the moon's surface. These facts resolved the dispute between the hot-moon and cold-moon scientists. The moon had indeed been a hot planet in its youth, but the episode of volcanism was brief; the moon died, volcanically speaking, three billion years ago, and has been dead and cold ever since. All the lava flows on its surface have turned out to be very old, as the cold-moon theorists had said they would be; but because the moon's surface changes so little with time, those ancient lava flows look as though they had happened yesterday. This fact misled the hot-moon scientists into believing that the moon was still active.

An important question remains unanswered: why did the moon's fires die out three billion years ago, while the earth is still volcanically alive? The answer has to do with the size of the moon. Small planets, like small animals, lose their heat to space quickly. The moon melted for a brief period near the beginning of its existence because radioactive elements heated its interior;* but then, as the radioactive heat passed quickly to the surface, it cooled off and became largely solid again. The earth, being a larger planet, held in its radioactive heat better, and has remained hot and partly molten down to the present time. Eventually, the earth will also become cold and volcanically inactive, but that will not happen for several billion years.

The discovery that the moon had active volcanoes at one time in its youth tells us something important; large amounts of water may

---

*The Apollo astronauts confirmed that the moon contains uranium and other radioactive substances. When these radioactive substances decay, they release small packets of energy which heat the rock in the interior of a planet and cause it to melt.

once have existed on the moon. Volcanoes are believed to be the source of most of the water on the earth, or on any planet. Bubbles of water vapor, trapped in molten rock, are released at the surface of the planet when its volcanoes erupt. When the vapor cools and condenses, the water flows into hollows in the crust to form the planet's oceans.

If water was indeed present, it follows that the moon may once have harbored life. Water is the quintessential element for the evolution of life; it provides a fluid medium in which the molecular building blocks of life can move freely. This free movement leads to frequent collisions between neighboring molecules; and the collisions lead to the chemical reactions which are the first steps in the evolution of the living organism.

Even if life evolved on the moon during that volcanically active period, its descendants could not have survived down to the present time; for when the lunar volcanoes died out the water would have leaked away to space, leaving the moon bone-dry. But the remains of this life may still be preserved somewhere. Where shall we look for them? Not on the moon's surface, because the sun's ultraviolet radiation, beating down relentlessly for billions of years, must long since have destroyed the traces of any organisms that once lived on the moon. But buried at some depth, protected against the solar ultraviolet rays, there we may find a record of ancient lunar life.

Whatever discoveries await us in the future exploration of the moon, the first lunar landings have already yielded the most extraordinary result that could have been hoped for. We are now certain that some of the materials on the moon have lain there since the beginning of time in our solar system. There is no longer any doubt that the record of the distant past, missing on the earth, can be deciphered on the airless, waterless moon. The moon is the Rosetta Stone of the planets.

EXPLORATION
OF THE MOON

The Ancient Surface of the Moon

The First Landing

"The Surface is Fine and Powdery"

A Typical Scene on the Lunar Seas

An Important Apollo Discovery

Oceans of Lava

The Oldest Region on the Moon

THE ANCIENT SURFACE OF THE MOON. The pitted landscape of the moon, carrying the scars of tens of thousands of meteorite collisions, bears witness to the excellect state of preservation of the lunar surface. The largest craters are visible in the photograph at left, made by joining two separate pictures of the half moon. The area outlined in white is shown in greater detail in the photograph at right, taken with the 100-inch telescope on Mount Wilson.

Meteorites have produced similar craters on the earth throughout its history, but most of these have been erased by the wearing action of wind and running water. Only the most recently formed craters, such as the Arizona Meteorite Crater (below), are still visible. The Arizona Crater, approximately one mile in diameter, is very similar in shape to craters of the same size on the moon. It was formed in the recent past, about 30,000 years ago, and will have vanished in 10 million years. Similar craters on the moon last for billions of years because the moon has no water or air and very little erosion.

THE FIRST LANDING. Eagle touched down on the surface of the Sea of Tranquility on July 20, 1969, at 4:18 p.m. EDT. The photograph at left, taken by Neil Armstrong, shows Edwin Aldrin emerging from the LM thirty minutes after the landing. He appears to float down the ladder because the force of gravity on the moon is one-sixth of the earth's gravity.

The reduced force of lunar gravity also explains the surprising tilts of the astronauts' bodies as they loped over the surface. Because they weighed so little, the friction between their boots and the ground was diminished. To gain the extra friction needed for a change in course, they had to lean into the direction they wished to follow and push against the ground at an exaggerated angle. Aldrin reported later, "Like a football player, you just have to split out to the side and cut a little bit."

"THE SURFACE IS FINE AND POWDERY," Armstrong said. "I can pick it up loosely with my toe. It does adhere in fine layers like powdered charcoal to the sole and sides of my boots. I only go in a small fraction of an inch. Maybe an inch, but I can see the footprints of my boots and the treads in the fine sandy particles." Armstrong's boots sank in six or seven inches in some places. The Armstrong footprint in the photograph *below* will remain clearly visible for ten million years.

A TYPICAL SCENE ON THE LUNAR SEAS. Most of the surface of the moon resembles this monotonous terrain, studded with small craters and littered with rock fragments. The powdery soil is heavily trampled by the two hour exploration. The landscapes lack color, being variously reported by the astronauts as a uniform dark gray, or sometimes gray with a hint of a warmer tone. But the astronauts were fascinated by the harsh contrasts of light and shadow, unsoftened by an atmosphere. Armstrong reported, "It has a stark beauty all its own. It's like much of the high desert of the United States."

The photograph, taken by Neil Armstrong during the first lunar landing, shows Aldrin setting up the solar wind experiment, in which a sheet of aluminum foil is placed to catch fast-moving particles blown off the surface of the sun. Because of their high speed, the particles penetrate the outer layers of the foil and are trapped within it. The foil is rolled up like a window shade, carried back to the earth, and heated in the laboratory to dislodge the trapped particles. The experiment yielded direct information on the chemical elements in the body of the sun.

AN IMPORTANT APOLLO DISCOVERY. The instrument shown in this photograph is a seismometer, placed on the moon by the astronauts to measure moonquake activity. The panels on either side of the seismometer are solar cells, which generate electricity from sunlight to operate the siesmometer.

The Apollo seismometers continued to work until 1977 — seven years after the last astronaut left the moon — radioing their data back to the NASA Center in Houston. They revealed that quakes occur far less often on the moon than on the earth, the most powerful moonquakes detected by the lunar seismometers having a rating of 2 on the Richter scale. If you were standing directly over a quake of this size on the earth, it would not produce a perceptible vibration in your feet.

OCEANS OF LAVA. The low level of moonquake activity measured by the Apollo seismometers (p. 112) proves that the moon is geologically dead today, with no active volcanoes. However, there are signs that the moon was volcanically active for a period in its youth. The lunar seas, such as the Ocean of Storms *(opposite)*, are composed of congealed lava that welled up from the moon's interior as molten rock several billion years ago and collected in the basins of the moon's crust, just as water collects in the basins of the earth's crust to form our oceans.

The cavities in the moon rock *(below)* prove that volcanoes existed on the moon at one time. These cavities are of a kind formed in lava when it cools rapidly and solidifies, trapping bubbles of volcanic gas.

All the lava-like rocks from the lunar seas turned out to be three billion years old or older, indicating that the moon died, volcanically speaking, three billion years ago, and has been dead since that time. Since volcanoes are the source of water and atmospheric gases on a planet, this finding explains the dry, airless state of the moon.

THE OLDEST REGION ON THE MOON. The lunar highlands *(opposite)*, the oldest part of the moon, are an exceptionally rough and heavily cratered terrain. In the photograph, Harrison Schmitt *(lower left)*, later a U.S. Senator, is inspecting a house-sized rock at the edge of the highlands, during the final Apollo mission. The rock was probably blasted out of the highlands several hundred million years ago, when a meteorite collided with the moon and excavated a large crater.

Some parts of the highlands contain rocks that were formed when the moon solidified at the time of its birth. These rocks are 4.6 billion years old — as old as the solar system itself. No region on the surface of the earth is as ancient as this.

9 VENUS

Nine planets circle the sun. Six of these—Mercury, Venus, Earth, Mars, Jupiter and Saturn—were known to the Ancients, while three—Uranus, Neptune and Pluto—were discovered in modern times. Mercury is the closest of the nine to the sun. It is a small planet, less than half the size of the earth, and scarcely larger than the moon. Its rocky, barren surface, alternately baked on the side facing the sun and frozen on the side facing away, is extremely inhospitable to life. The planet is difficult to reach by rocket from the earth, because of its closeness to the sun, and it is unlikely that we will learn more about it than we know for many years to come.

Moving outward from the sun beyond Mercury, we come to Venus. Venus is the earth's closest planetary neighbor. It is also our sister planet, closely similar to the earth in size and weight, and situated at a distance from the sun which is not very different. The surface of Venus is completely covered by clouds, and conditions on the planet have always been an enigma, yet romantic hope has flourished that beneath these clouds lie teeming masses of flora and fauna. In 1686 de Fontenelle, in his book *Conversation on the Plurality of Worlds*, described the characteristics he expected to find in the people on Venus:

> —I can tell from here . . . what the inhabitants of Venus are like; they resemble the Moors of Granada; a small black people, burned by the sun, full of wit and fire, always in love, arranging festivals, dances and tournaments every day.

In fact, Venus should provide an even more agreeable climate for living organisms than the earth. The planet is 70 million miles from the sun, while the distance of the earth is 93 million miles. Because Venus is closer it receives twice the intensity of the sunlight falling on the earth; and although its heavy cloud cover keeps out some of this solar energy, we can still estimate that on Venus the average temperature at the latitude of London should be a comfortable 80 degrees Fahrenheit, or approximately the same as the balmy temperatures of the islands of the Caribbean.

But some years ago, evidence came to light which indicated that the climate of Venus is far from balmy. Measurements of the intensity of the radiation emitted from the planet suggested that the temperature on its surface is 1000 degrees Fahrenheit, which is hot enough to melt lead. It is also hot enough to break apart all the delicate molecules that make up the ingredients of a living cell. No organism remotely resembling terrestrial life could survive in such heat.

Yet hope lingered on for the discovery of a green world on Venus. Some astronomers argued that life might be supported at the north and south poles, which should be cooler. Others suggested that bizarre organisms might have developed on Venus—gas-filled bladders, perhaps, whose buoyancy would cause them to float high in the atmosphere, where the air is considerably cooler than it is near the ground.

But in the last few years, several Russian and American spacecraft journeyed to Venus and carried out measurements that have removed the last trace of doubt regarding the high temperature on its surface. The spacecraft measured the temperature on the planet in several different ways, but their data led to the same conclusion. Venus is indeed hot enough to melt lead, and there is no reasonable chance of finding life on its surface.

The spacecraft revealed other unpleasant facts about Venus. The planet has almost no water, and what little it has exists in the form of superheated steam;* the fleecy, white clouds are droplets of deadly

---

*The dryness of Venus is a puzzle to students of the planets. According to current views, Venus and the earth condensed out of the same materials, containing similar amounts of water. The water on the earth, if spread uniformly over the surface of

sulphuric acid; and the atmosphere is an asphyxiating blanket of carbon dioxide, so dense that its pressure would exert a force of ten thousand tons on the human body. No flesh-and-blood creature could stand the conditions on Venus. The planet is lifeless today and will remain lifeless in the future; its hellish surface is beyond our reach.

It is surprising that Venus is so much hotter than the earth. Why did two planets, probably formed out of similar materials and situated at comparable distances from the sun, evolve along different paths? Why is the surface of Venus baked by a searing heat, while the earth luxuriates in a climate friendly to all known forms of life? The Soviet and American spacecraft carried out measurements which supplied the answer to this question. According to the spacecraft measurements, the atmosphere of Venus consists primarily of a heavy layer of carbon dioxide, about 100,000 times more than is in the atmosphere of the earth. This dense atmosphere of carbon dioxide acts as an insulating blanket which seals in the planet's heat and prevents it from escaping to space. The trapped heat raises the surface to a higher temperature than it would have if the planet were a naked, airless body of rock. Calculations based on the insulating properties of carbon dioxide show that the temperature of Venus could easily be raised to 1000 degrees as a result of this effect.

The high temperature on Venus is explained by the abundance of carbon dioxide in its atmosphere. If the earth had as much carbon dioxide, the temperature on our planet would also be unbearable. Why do we have less carbon dioxide than Venus?

To some extent, the answer is connected with the presence of life on the earth. At the present time, much of the carbon dioxide in the earth's atmosphere is removed by marine animals, which absorb this

the planet, would form a layer about 8000 feet deep. A layer of water of approximately the same thickness should also exist on Venus. Because of the high temperature on Venus' surface, the water would be present not in liquid form, but in the form of water vapor, or steam, in the atmosphere. The Soviet spacecraft showed that nearly all this water is missing; the amount present in the form of steam is equivalent to a layer of liquid only one foot thick.

gas in sea water and convert it within their bodies to solid substances known as carbonates. Sea shells, for example, are nearly pure calcium carbonate. Today the upper layers of the earth's crust contain a thick layer of carbonate, formed by the compressed shells of countless mollusks and crustaceans that died long ago; and locked up in these carbonates is a large part of the carbon dioxide that would otherwise blight our atmosphere, as it blights the atmosphere of our less fortunate neighbor planet.

When the earth was young, life was either absent or scarce, and carbon dioxide could not have been removed in this way. However, the gas still could have been absorbed from the atmosphere of the young earth by other chemical changes, not involving living organisms. In these reactions, atmospheric carbon dioxide combines with rocks to form carbonates, somewhat as oxygen in the atmosphere combines with iron to form rust. Such reactions do not take place at an appreciable rate if the rocks are hot and dry. Therefore, they cannot occur on Venus. But they do occur on the earth, because it is cooler and more moist.

In other words, carbon dioxide is removed from the atmosphere of the earth in many ways but cannot be removed in any way from the atmosphere of Venus. It is understandable that Venus now has a great abundance of this gas in its atmosphere.

The present concentration of carbon dioxide on Venus must have built up slowly in the atmosphere over hundreds of millions of years. Why did life not develop in this initial period, before too much of the harmful, heat-insulating gas accumulated? If life had once spread over Venus, its ability to absorb carbon dioxide could have kept the planet comfortable forever after. The answer is that Venus, being somewhat closer to the sun, was slightly warmer than the earth when both planets were young and still lifeless. As volcanoes erupted on the two planets, carbon dioxide began to accumulate in their atmospheres. Because of the moderately higher temperatures on Venus, the rocks on that planet did not absorb the gas quite as effectively as on the earth. The concentration of carbon dioxide continued to build up on Venus, sealing in the planet's heat and making its surface still hotter. The hotter the surface became, the less carbon dioxide it could absorb. The curve of temperature growth spiralled upward, until finally the conditions needed for the development of life on Venus were permanently destroyed.

How far from the sun must a planet be to maintain a comfortable temperature for life? As yet, we do not know; the spacecraft results tell us only that Venus was too close, and that was its undoing. If it had been only a few million miles farther away, the temperature of its surface might have climbed slowly enough to permit life to gain a toehold; and once life began, it could have held the abundance of carbon dioxide in check, and prevented the temperature from climbing out of bounds subsequently. Venus, having lost its chance to harbor life when the solar system was first formed, could never again recapture the opportunity.

10 MARS

Mars has generated more speculation regarding extraterrestrial life than any other planet in the solar system. Several properties of the planet have contributed to these speculations. Its surface, nearly free of clouds, reveals changes during the Martian year that resemble the march of the seasons on the earth. In each hemisphere a polar cap grows larger in the fall and winter and diminishes in the spring and summer. Dark regions appear each spring that are suggestive of the seasonal growth of vegetation. Toward the end of the nineteenth century, some observers reported a planetary network of canals presumably engineered by intelligent life.

Twentieth-century studies of Mars have failed to confirm the existence of the canals. However, spacecraft observations made in recent years have contributed a great deal of new information that bears directly on the prospects for Martian life.

The density of Mars, like that of Venus, is about the same as the density of the rocks lying on the surface of the earth; for this reason Mars is believed to be composed of rocky materials similar to those on our planet. The atmosphere of Mars is rather thin, perhaps a hundredth as dense as the earth's atmosphere. On Mars, unlike Venus, only a trace of clouds exists, and the features of the Martian surface are not appreciably obscured by them. Dust storms and haze occasionally are conspicuous, but most of the time the planet is open to photographic surveillance.

In spite of the thinness of the Martian air, it is impossible to obtain good photographs of Mars from telescopes on the earth because of the blurring effect of the earth's atmosphere on rays of light reaching us from it. No features of the surface of Mars can be seen from the earth, no matter how large the telescope in which the planet is viewed unless they are 50 miles or more in diameter. It is impossible to tell from the earth whether Mars has mountains, ocean beds, or any features that might indicate the presence of life.

Far better pictures of Mars have been taken by NASA spacecraft as they swept past the planet at long distances. The clearest of the spacecraft photographs revealed features as small as a few feet across. They showed that the surface of Mars is marked by a large number of craters, presumably produced by meteorite collisions. There are relatively fewer craters than on the moon, but far more than on the earth.

The spacecraft photographs suggest that Mars stands midway between the moon and the earth as a planetary body, its surface being older and better preserved than the surface of the earth, but not as well preserved as the surface of the moon.

Mars also stands midway between the moon and the earth in size and mass. It is approximately twice the diameter of the moon, and one-half the diameter of the earth; its mass is approximately 10 times the mass of the moon and one-tenth the mass of the earth. These facts lead to a prediction regarding volcanism on Mars. We know from the study of the earth that volcanic eruptions result from the release of heat within the body of a planet. The source of the heat is the decay of radioactive elements, which exist in the interior of the earth in small concentrations. The findings from the lunar landings indicate that these radioactive elements also exist in the moon. Presumably they exist in the interior of Mars as well.

The heat that the radioactive elements created must have gradually increased the temperatures in the interiors of the moon and Mars, as well as the earth, during the early years of their existence. Volcanic eruptions would have commenced on the three planetary bodies at about the same time, when they were about a billion years old, as a result of the internal heating. The eruptions must have released gases that would create atmospheres on these planets; they must also have released copious amounts of water vapor, that would have accumulated to form oceans on their surfaces.

Although volcanism probably began at about the same time on the earth, the moon, and Mars, it did not last for the same length of time on each because of the differences in their sizes. The reason is that the heat generated in the interior of a small planet has to travel only a short distance to reach the surface. Thus, the heat escapes quickly, and volcanism terminates. This appears to have been the case for the moon, whose episode of volcanism lasted only about one billion years. Since volcanoes bring air and water to a planet's surface, the moon may have had an atmosphere and oceans during its brief period of volcanic activity. But after its volcanoes became extinct, the air and water molecules would have leaked away to space.*

If a planet is large, its internal heat must travel through a thicker layer of rock to reach the surface. The layer of rock acts as an insulating blanket, bottling up the radioactive heat and causing the temperature in the interior to remain at a high level for a relatively long time. This is the case for the earth, which has been volcanically active throughout most of its lifetime, and is still active today. That explains the fact that the earth has a great deal of air and water.

Since Mars is intermediate in size between the moon and the earth, it must have retained its radioactive heat longer than the moon, but not as long as the earth. Therefore Mars should have been volcanically active for longer than a billion years, but may not still be volcanically active today.

If volcanoes did persist on Mars for more than a billion years, the surface of Mars must have once been covered by water to a considerable depth. A substantial atmosphere also existed on Mars during this long period of volcanic activity. These circumstances would have created agreeable conditions for the evolution of life.

The prediction that Mars had volcanoes was confirmed in 1976, when television cameras on the Viking spacecraft obtained excellent pictures of the entire planet. When the spacecraft first reached Mars, the planet was obscured by a violent dust storm that lasted for nearly

---

*Because the moon is a small object with a weak pull of gravity, it must have lost its molecules of air and water to space very rapidly.

two months. Eventually the dust settled, and the cameras revealed details of the surface of Mars that had never before been seen by man. The most conspicuous feature in the photographs was a huge volcanic mountain, later named Mount Olympus. This mammoth volcano is fifteen miles high and three hundred miles across at its base.

In every respect Mount Olympus resembles the mounds of congealed lava that form on the earth when successive outpourings of molten rock occur at a single spot over a period of millions of years. If the Pacific Ocean basin could be emptied, the Hawaiian Islands would be revealed as similar but smaller mounds of lava, rising out of the floor of the ocean in the same way that the Mars mountain rises out of the surrounding terrain.

About a dozen large volcanic mountains like Mount Olympus have been discovered on Mars. There is evidence that all are extinct, and have been for some time. Their flanks have many small meteorite craters, which would have been obliterated if fresh floods of lava had run down the sides of the mountains in recent years. From the number of craters present, it is estimated that the Martian volcanoes died out about half a billion years ago, after about three billion years of volcanism. This fact agrees with the prediction that Mars should have been volcanically active longer than the moon, but not as long as the earth.

The existence of volcanoes on Mars suggests that water was plentiful on Mars at one time. The spacecraft photographs also confirm this prediction. They reveal dry channels that look like arroyos and riverbeds carved by flash floods that occurred millions of years ago. Nothing except the flow of enormous volumes of water could have produced those channels.

Furthermore, the north and south poles of Mars are covered by thick caps of ice that show signs of having been partly melted and refrozen several times in the past. The implication is that Mars has experienced a succession of ice ages alternating with times of warmth and moisture. Not only has Mars been wetter in the past; it has also been warmer.

No developments more encouraging for the prospect of Martian life could be imagined. If life gained a toehold on Mars during those warm, wet intervals, it might have adapted by degrees to the harsh conditions that befell the planet later on. The transition to the drier climate of today could have occurred very slowly, over a period of millions of years and a like number of generations. During this long period of slowly increasing aridity, the weakest individuals in each generation would be eliminated, and the hardiest would remain, propagating their qualities of strength to their descendants. There seems no reason to doubt that varied and interesting forms could exist on Mars today as a result of this long-continued process of natural selection, *if* the planet once had an abundance of water. Of course, Martian life would be simple and perhaps no higher than the level of a microbe; there are surely no intelligent Martians, because all evolutionary progress on Mars must have been slowed down hundreds of millions of years ago, when the planet's volcanoes became extinct and its water supply diminished.

The search for life on Mars was planned with this last fact in mind. The Viking spacecraft contained three experiments that should have worked if any kind of living organism existed in the Martian soil, no matter how primitive. The first experiment was designed to test for gases given off by living organisms; for example, Martian plants would release oxygen as a waste product, and animals would release carbon dioxide. The second experiment tested the soil for plants or plantlike organisms; their presence would be revealed by the chemical reactions characteristic of a plant, in which sunlight, air and water are combined to form the ingredients of a living cell. The third experiment was designed to reveal Martian microbes; it worked by feeding the hypothetical microbes a special food that contained a very small amount of radioactive carbon; if microbes were present in the soil, they would ingest the food and exhale radioactive carbon dioxide, which would be detected by a Geiger counter in an adjoining chamber.

All three tests yielded positive results. However, a detailed study of the findings suggested that the results from the first two tests could be explained by chemical reactions in the soil rather than biological organisms. But the positive results from the microbe test were not easily explained in this way. In fact, the test survived every one of the checks and cross-checks devised to distinguish genuine life from chemical reactions. For example, in one check the soil samples used in the three tests were heated to 120 degrees Fahrenheit. This temperature increase would not affect most chemical reactions, but it would make bacteria feel distinctly uncomfortable, and their population would diminish. When the check was run, the microbe signal diminished. This result supported the interpretation that there must be genuine microbes in the soil.

Still, the Viking biologists have not been inclined to accept the microbe test as convincing evidence of Martian life. One reason for their caution is a result obtained from another instrument in the Viking spacecraft. This instrument tested the soil for the presence of organic molecules—the building blocks of living matter. It detected none. How can life exist in the Martian soil if the building blocks of life are not present?

The answer may be connected with the fact that the sensitivity of the organic-molecule detector was limited; it could detect microbes only if they were present in very large numbers, equivalent to the microbial content of a rich garden soil. If the Martian soil had contained a sparse population of living organisms, similar to the population of microbes in the soil from the Antarctic, the molecule detector would have missed them entirely.

Someday more elaborate tests for life will be carried out on Mars, either by men or by instruments, and the question will be settled. If Martian organisms are ever detected, we will know that life has arisen independently on two planets—the earth and Mars—in one solar system. Therefore, it cannot be a rare event; in fact, the chance of life arising on any of the billions of earthlike planets in the cosmos must be quite high. From this fact it must follow that the universe is teeming with creatures of every shape, size and level of development. This would be the most portentous discovery yielded by the exploration of space to date.

EXPLORATION
OF MARS

TWO VIEWS OF MARS. Photographs of Mars taken from the earth are blurred by the movement of currents of air in our atmosphere *(below)*.

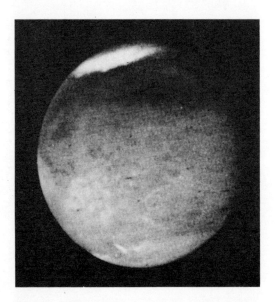

The first clear photographs were taken from a spacecraft approaching the planet *(opposite)*, and yielded a remarkable finding: Mars, unlike the moon, has mammoth volcanoes, some far larger than any volcano on the earth. The Martian volcanoes are clearly visible as dimples near the bottom of the spacecraft photograph.

Volcanoes are the source of water on a planet. Pockets of water, trapped in the interior of the planet, come to the surface as steam in volcanic eruptions, and condense to form the oceans. The Martian volcanoes suggest that Mars may once have had more water on its surface, and a favorable climate for living organisms.

MARTIAN VOLCANOES. The largest volcano on Mars is Mount Olympus *(opposite)*, a huge mound of congealed lava, 70,000 feet high and 300 miles across at the base — twice as high and twice as large as the biggest volcanoes on the earth. The volcano is extinct; its last eruption occurred about a billion years ago.

When Mount Olympus and other Mars volcanoes were active, they must have poured water vapor and other volcanic gases into the Martian atmosphere, creating a warmer, moister and more earthlike climate. Some of that water is still on Mars, locked up in the polar caps in the form of ice. From the size of craters partly buried in the ice, the caps appear to be hundreds of feet thick. If they were melted, the water released would be sufficient to cover the surface of Mars with a shallow sea. See also pages 174-175.

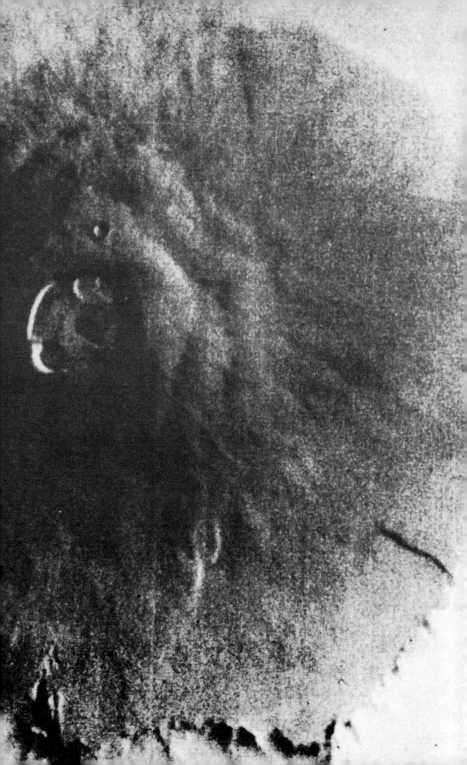

EVIDENCE FOR MARTIAN RIVERS. Photographs taken by the Viking spacecraft provide additional evidence that volumes of water once coursed across the Martian plains. The dry riverbed *(opposite)*, meandering across a gently sloping terrain, has a pattern of meanders *(arrow)* resembling the meanders in rivers on the earth, such as the Red River in Louisiana *(below)*, photographed from a NASA spacecraft in orbit. More evidence of Martian rivers appears on pages 138-143.

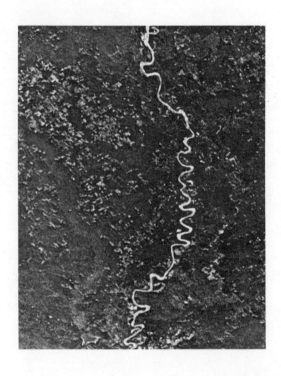

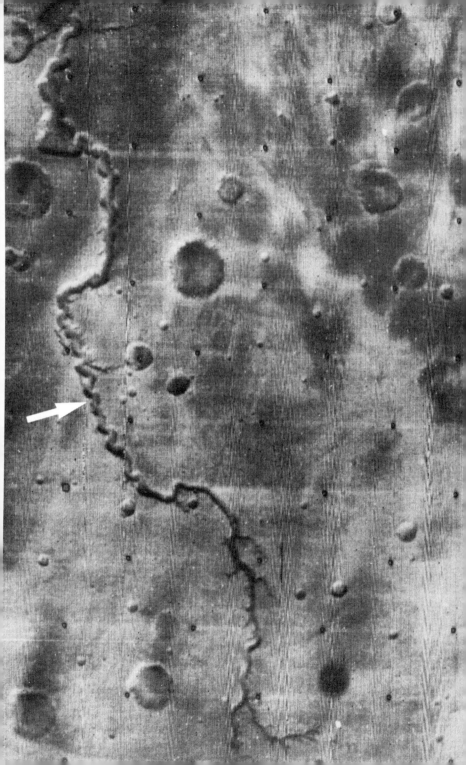

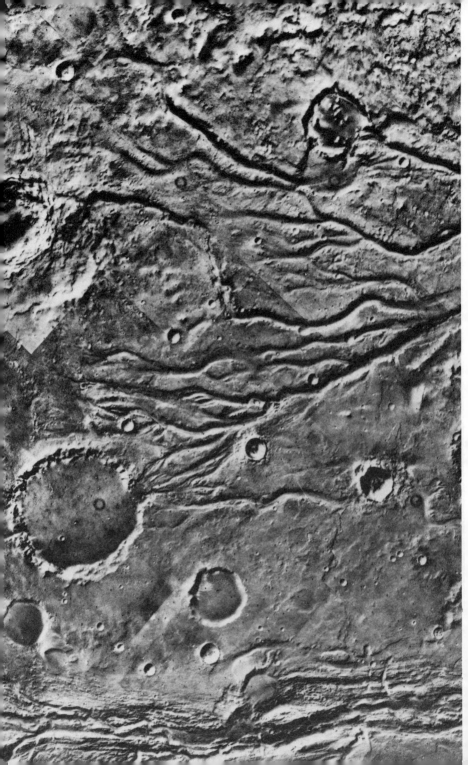

The photograph *(opposite)* shows converging channels that apparently carried the runoff from a major flood. The source of the water may have been the sudden melting of subsurface ice or permafrost. The earth photograph *(below)* shows similar channels created by water erosion in the Arizona desert.

The Martian channels *(opposite)* have "islands" *(arrow)* like those seen in rivers on the earth when water flows around a midstream obstacle. The characteristic teardrop shape of a midstream island appears even more clearly in the photograph *(below)*.

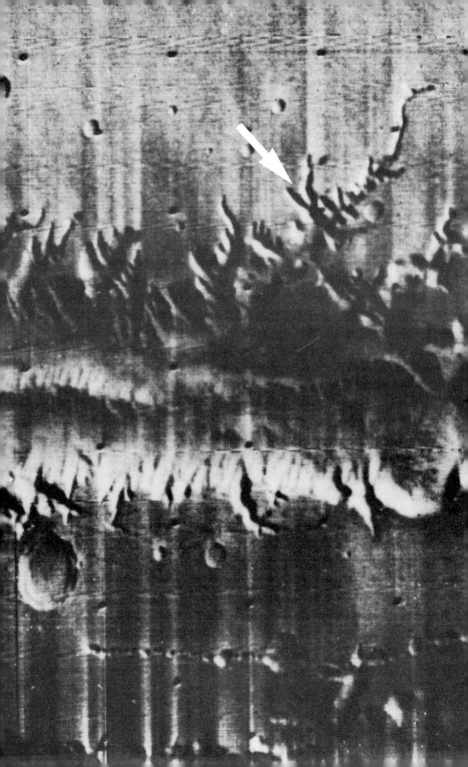

This Viking photograph shows a portion of the Valley of the Mariner, a 3000-mile-long crack running across the surface of Mars. Converging channels in the north (upper) rim *(arrow)* look like channels on the rim of the Grand Canyon *(below)*, carved by water pouring over the lip of the canyon.

THE POLAR CAPS. Because Mars is cold, its water is in the form of ice in the polar caps *(below)*, and possibly buried ice and permafrost elsewhere on the planet.

A detailed photograph of the edge of the polar cap *(below)* shows distinct layers, suggesting that Mars has had cold periods, or ice ages, alternating with times of warmer climate. The thickness of the layers indicates that the cold periods recur every million years or so. According to this interpretation, Mars is currently in the grip of an ice age; it has been warmer in the past and may become warmer in the future.

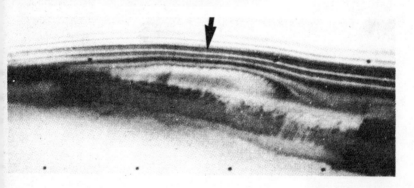

The photograph *opposite* shows tongues of ice — Martian glaciers — several miles long, advancing over the edge of the cap and into the region beyond. It confirms the picture of Mars as a planet experiencing an ice age.

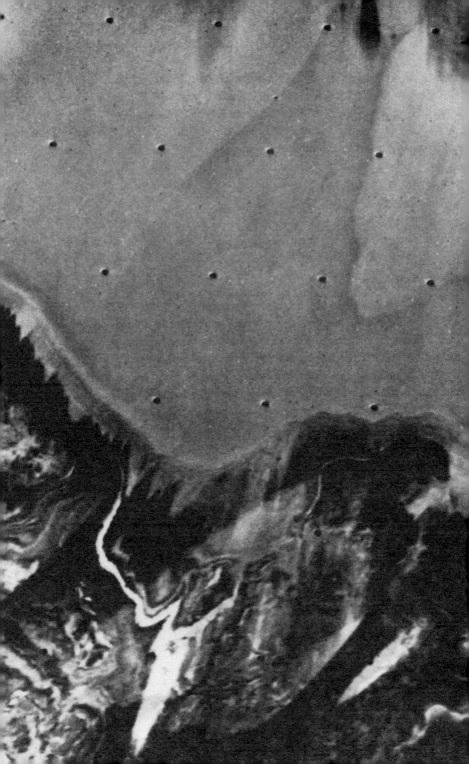

PROSPECTS FOR LIFE. A terrestrial animal called the water bear *(below, left)* provides an example of the extremes to which life might adapt on a dry planet like Mars if sufficient time were available. The water bear, a little animal about one twentieth of an inch in size, can survive for 100 years dehydrated to three percent of its normal body weight of water. Moistened, ...

it springs to life *(below, right).*

Any life that exists on Mars evolved long ago, during a period when the Martian volcanoes were still active and water was plentiful. If the water leaked away slowly enough, evolution on Mars could have produced even more extreme adaptations to sustained drought than the terrestrial water bear.

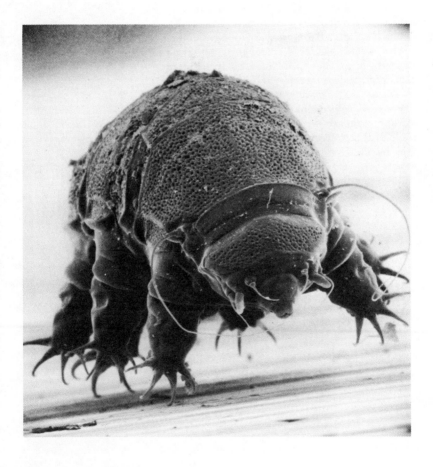

THE FIRST MARS LANDING. The search for life on Mars began in 1976, when the unmanned Viking spacecraft settled into an orbit around Mars and sent a landing craft *(opposite)* down to the surface of the planet for reconnaissance. The lander came to rest within a few miles of its intended location, an accuracy, relative to the distance travelled, equivalent to putting a bullet through a playing card 3000 miles away.

The Viking lander is twelve feet across and eight feet high. It resembles an oversized insect, with protruding eyes to either side of a long proboscis *(arrow)* that can be extended to pick up samples of soil. In operation, the proboscis draws in the sample and deposits it in the maw of the lander, where it is digested and analysed for signs of life.

The Viking lander has roughly the brain capacity of an insect, provided by an onboard computer. Unlike insects, it also has the ability to learn, i.e. to modify its behavior in accordance with instructions radioed from the Jet Propulsion Laboratory in California. This non-biological organism, made of metal and plastic, may be the forerunner of a race of intelligent robots with narrow interests but quasi-human capabilities, that will serve as surrogates for man in the exploration of space.

THE SEARCH FOR MARTIAN LIFE. The Viking proboscis, on command from the earth, pushed aside a Martian rock *(left panel, arrow)* to expose the soil underneath, which had been protected from the sun's ultraviolet radiation. This radiation is thought to create exotic chemical compounds in the Mars soil. Most of the results obtained from the Viking life tests can be explained by chemical reactions involving these compounds, rather than by living organisms.

The soil from under the rock, lacking the exotic compounds, should have yielded no "life" signal when it was tested. However, one test — for microbes — produced as strong a signal from the soil under the rock as it had from the samples on the surface. This finding contradicts the explanation of the Viking results in terms of chemical reactions. It suggests that the microbe test detected genuine microscopic organisms on Mars, and not merely an unusual chemical in the soil.

INTELLIGENT MARTIANS. In 1887, the Italian astronomer Giovanni Schiaparelli reported a network of canals on Mars. His report was subsequently confirmed by other astronomers, and Percival Lowell described the canals as "a vision of a thread stretched across orange seas." However, not all astronomers saw them. In the drawings of Mars *opposite*, the one *above*, made by Schiaparelli on the basis of telescope sightings, shows many canals clearly marked. The drawing *below*, showing the same region on the planet as observed by E. M. Antoniadi, has similar markings but no hint of canals.

The canals, although sighted visually through telescopes, never appeared on photographs from the earth. Recent close-ups of Mars from passing spacecraft also show no signs of them. It is now generally believed that they were imagined by hopeful astronomers straining at the limits of visibility.

There is no possibility of advanced or intelligent life on Mars, although the Viking measurements hint at the presence of simple, microscopic organisms. Evolution on Mars must have halted or slowed down at least a billion years ago, when the Martian volcanoes — source of water on the planet — became extinct. At that time, life on Mars would only have been at the stage of life on the earth one billion years ago, when the most advanced organism here was a creature resembling a worm.

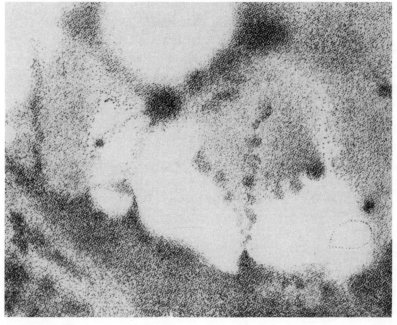

what is seen
in photos
Frozen
ammonia

H₂O

clouds ←— sea of Hydrogen
40,000 miles thick

11 JUPITER: CRUCIBLE OF LIFE

IO - is alive - possible life form

Five planets lie outside the orbit of Mars. These are the giant planets—Jupiter, Saturn, Uranus and Neptune—and the small planet Pluto.

Pluto, found in 1930, was the ninth and last planet to be discovered in the solar system. Its orbit is farther from the sun than that of any other planet and probably marks the outer boundary of the solar system. Because Pluto is so far away, we have been able to learn very little about it, except that it appears to be a body similar in size and composition to the earth. It must be a frozen, silent world, far too cold to support any form of life.

More is known about the giant planets. They are approximately 10 times larger than the earth and 100 times more massive, but considerably lower in density. In general, their density is about the same as that of water; Saturn, in fact, is less dense than water; it would float in the bathtub if you could get it in.

The giant planets are less dense than the earth and its neighbors because they are made up largely of the lightest elements, hydrogen and helium. These elements make up most of the matter in the universe; they also constitute most of the matter in the sun and in the giant planets; but for some reason not clearly understood, they are missing from the earth and inner planets. Perhaps the particles and radiation emitted from the sun in its early years blasted the hydrogen and helium out of the inner parts of the solar system, from which the earth was formed, while the outer regions, out of which the giant planets were formed, were too far away to be affected by the blasting action.

Jupiter is the largest of the giant planets and the most massive planet in the solar system. It is a huge, rapidly spinning ball of liquid and gas 90,000 miles in diameter—more than 10 times the size of the earth—and 318 times as massive. Jupiter is larger than all the other planets combined. Because of its enormous mass, the gravity on Jupiter is much stronger than on the earth. An astronaut landing on the giant planet would be plastered to the spot by its gravitational force, which would give him a weight of 600 pounds. But the landing can never take place because Jupiter has no surface to stand on. Beneath its swirling clouds there is only a vast sea of molten hydrogen—hot, corrosive, destructive to life.

Jupiter is made mainly of hydrogen and helium. It also contains some iron and rock, enough to make up several earths. The rocky materials are probably concentrated at the center of Jupiter, forming a core several thousand miles in diameter. If the core were exposed, it would look like the kind of planet with which we are familiar. However, it is buried beneath a mantle of molten hydrogen 40,000 miles thick, which makes up the bulk of Jupiter.

The molten hydrogen merges imperceptibly into Jupiter's dense atmosphere. The atmosphere contains, in addition to hydrogen and helium, appreciable amounts of noxious gases such as ammonia, methane and acetylene. Water also occurs, in the form of steam in the lower atmosphere where the temperature is high, and there are droplets of liquid water at higher altitudes where the air is cooler.

The water droplets form a thick layer of clouds covering the entire planet. A second layer of clouds, composed of tiny crystals of frozen ammonia, floats above the water clouds. The clouds of ammonia form the visible face of Jupiter; it is these clouds that appear in the striking photographs of the planet taken by the spacecraft cameras.

A spacecraft named Voyager reached Jupiter in 1979, flew over the planet's surface at a distance of 170,000 miles, photographed the mysterious Red Spot, and reconnoitered the four earthlike moons of Jupiter before continuing on to a rendezvous with Saturn in 1980. Voyager photographs of Jupiter and its four largest moons—Io, Europa, Ganymede and Callisto—are included in the Color Plates.

The most extraordinary feature in the Voyager photographs of

Jupiter is the Red Spot, an orange-red oval 15,000 miles long and 8000 miles wide, roughly the size of the Pacific Ocean. The photographs reveal swirling motions in the vicinity of the Red Spot, which suggest that the Spot is a mammoth hurricane. However, the Red Spot has persisted for at least 300 years, whereas hurricanes and storms on our planet play themselves out and disappear after a few days at most, because their energy is dissipated into turbulence in the surrounding air. What force has fed energy into the Red Spot to keep it intact for centuries? No one knows.

There is another puzzle in the Spot: Hurricanes and storms tend to wander over the surface, whereas the Red Spot is anchored in one place. A huge object, lying beneath the clouds on Jupiter, must be the Red Spot's anchor. What is this object? That question is also unanswered. Although the Voyager photographs show the Red Spot in more detail than it has ever been seen before, it remains a mystery.

The Voyager made a major discovery as it swept past the planet. Jupiter, like Saturn, is surrounded by a huge ring of rock fragments, 160,000 miles in diameter. The existence of the ring had been un-suspected because, unlike Saturn's rings, it is too faint to be visible through earth telescopes. The Voyager cameras were so close that they couldn't miss it.

Although the discovery of Jupiter's ring was a surprise, its occur-rence is not difficult to explain with hindsight. Suppose you were in orbit around Jupiter with your feet pointed toward the planet. Since a planet's gravitational pull is stronger on objects that are closer to it, your feet would feel a stronger force toward Jupiter than your head. The result would be a tendency to separate your head from your feet, i.e., to pull you apart.

In the same way, a moon circling Jupiter must feel a stronger gravitational attraction on its near side, closest to the planet, than on its far side, tending to pull it apart. However, this tendency is resisted by the gravitational attraction of the moon itself, which tends to draw all parts of it together.

If the moon is at a considerable distance from Jupiter, its internal force of gravity, holding it together, will be stronger than the disruptive effect of Jupiter's gravity, and no damage will occur. But if the moon approaches too close to Jupiter, the planet's gravity will break it up into many small fragments. The fragments gradually spread out into a ring of debris circling the planet. This is the explanation for Saturn's rings, and presumably also the explanation for the ring around Jupiter.

How close could one of Jupiter's moons approach the planet without being broken up? A formula based on a comparison between the moon's gravity and Jupiter's gravity tells us that if the moon is more than 100,000 miles from the center of Jupiter, it is safe; if it is within that distance, it will be broken up. Since Jupiter's ring is about 80,000 miles from the center of the planet—well within the critical distance—its existence is accounted for nicely by the theory.

Just outside the critical distance for breakup into a ring, a small red satellite called Amalthea orbits Jupiter at a distance of 110,000 miles. Voyager captured a somewhat blurred image of Amalthea as the little moon raced around its parent planet at a speed of 70,000 miles per hour. The Voyager photograph reveals Amalthea to be a potato-shaped chunk of rock about 100 miles long and 80 miles wide. This little satellite may have been captured from the asteroid belt, or it may have condensed out of the cloud of gaseous dust surrounding Jupiter at the same time that the planet was forming.

Beyond Amalthea lies the satellite Io named after one of the mythological lovers of Jupiter. Io was also photographed by Voyager, and yielded the second big surprise of the mission. Io is alive. It has actively erupting volcanoes; in fact, one volcano erupted while the photographs were being taken.

Our moon is dead; it died three billion years ago. Mars is dead; its last volcanoes died out a billion years ago. Io and the earth are the only volcanically active planetary bodies known in the solar system. Why is this important? Because a planet with actively erupting volcanoes has water and warmth—two elements essential for life.

Volcanoes and life—the two are inextricably linked. A student of the evolution of planets and life feels a warmth in his heart as he contemplates the picture of an active volcano on Io. Io has no atmosphere in the ordinary sense, and no warm, shallow seas, conducive to nature's experiments with evolution; yet it is a friendly planet, more like the earth in some ways than any other we have visited.

The evidence for active volcanoes on Io is convincing, but it is very hard to understand why this moon is volcanic. Why is the interior of Io so hot? Io is a small body, only slightly larger than the earth's moon, and small moons and planets, like small animals, lose their internal heat quickly. That is why our moon's fires went out three billion years ago. Io's fires should also have died out around that time. Why do they still burn?

The answer appears to be connected with Jupiter's powerful gravity. Tugging at the near side of Io harder than the far side, the gravity of the Giant Planet causes the interior of the moon to yield somewhat, creating friction and heating. According to calculations, the heat accumulated from this friction, produced by the repetition of orbit after orbit over millions of years, would be sufficient to melt the interior of Io.

The effect is similar to the one that causes moons to break up into fragments and form rings. The difference is that Io, being a little farther away from Jupiter than the critical distance for forming a ring, was not broken up, but only twisted and pulled about somewhat.

Beyond Io lie three large moons and eight small ones. The small moons, probably captured from the asteroid belt, were not investigated on this Voyager mission. However, the three large moons were studied carefully. Their names, in order of increasing distance from the planet, are Europa, Ganymede and Callisto—all lovers of Jupiter in classical myth. They were examined, along with Io, by Galileo in 1610 when he first turned his telescope on Jupiter. Their relatively rapid motions around their parent planet suggested to Galileo that Jupiter was a solar system in miniature. He concluded that the earth could not be the center of all motion in the universe. This was the most powerful argument yet discovered for the theories of Copernicus. Jupiter's moons were responsible to a large degree for the trouble Galileo found himself in a few years later.

Ganymede, Callisto and Europa are known as the icy satellites

AN ALBUM
OF STARS
AND PLANETS

THE HORSEHEAD NEBULA. A dense cloud of gas and dust silhouetted against a bright background of light from young stars in the Milky Way. This nebula is located in one of the breeding grounds of new stars in our galaxy.

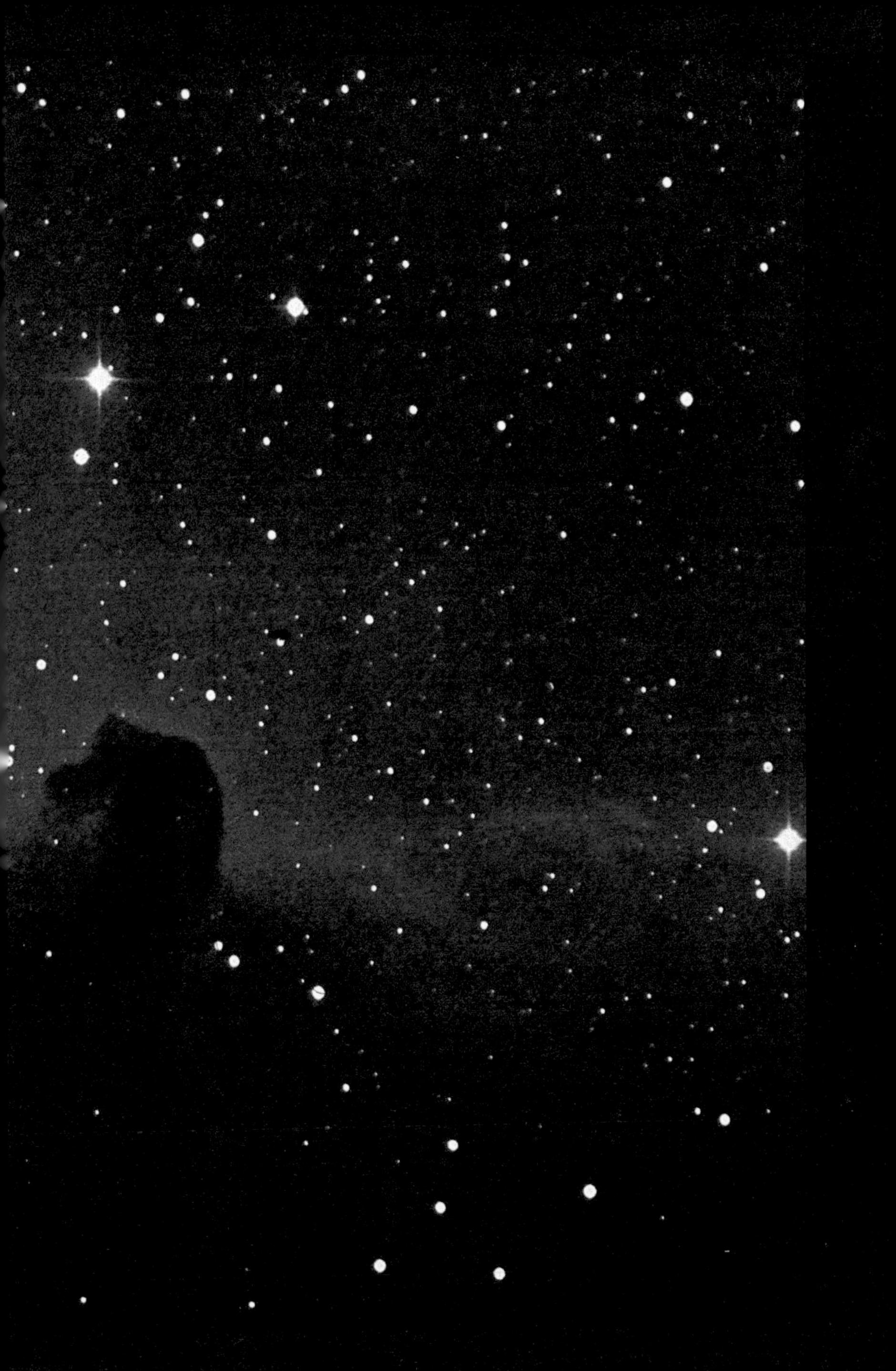

THE ORION NEBULA. This region is also a breeding ground of new stars. Most of the stars in Orion were formed only six to ten million years ago, in the period when man's ancestors had recently branched off from the apes in the forest and begun to walk erect. The extraordinary colors in the nebula are emitted by clouds of hydrogen atoms which absorb ultraviolet rays from the many newborn stars in their midst.

162

THE PLEIADES. A cluster of relatively young stars born 60 million years ago, shortly after the last dinosaurs vanished from the earth. The blue halo around each of the brightest stars is light from the star itself, reflected from grains of dust in interstellar space.

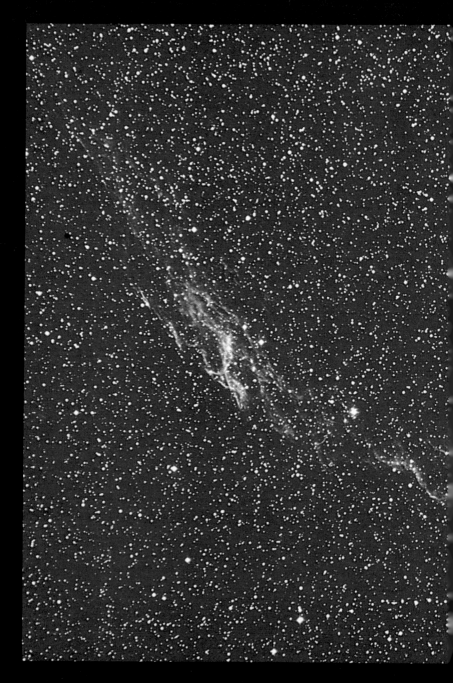

THE VEIL NEBULA. A remnant of an exploded star (pp. 74-75).

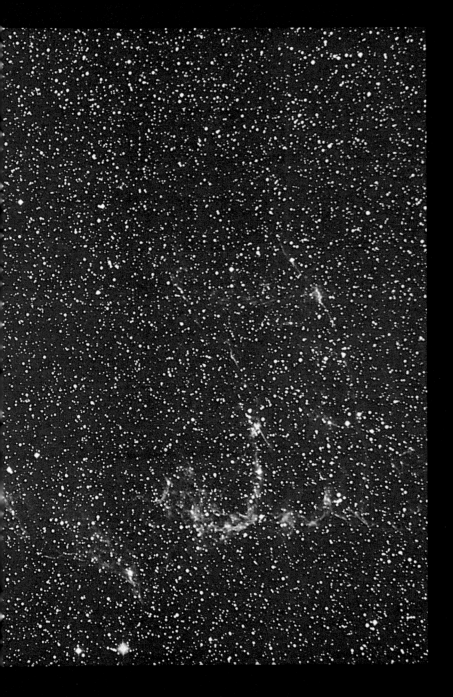

EARTHRISE. Modern astronomy was born in the realization that space is vast and the world of men is small.

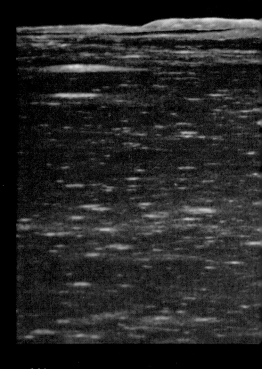

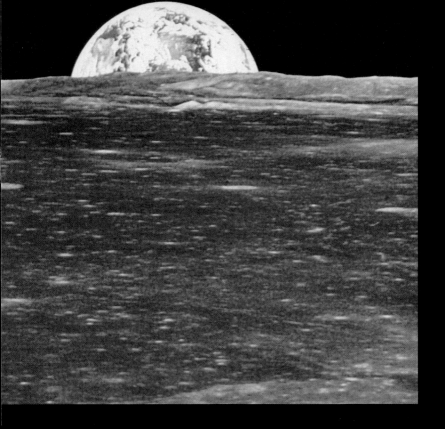

THE EARTH, MOON AND MARS PHOTOGRAPHED FROM SPACE. The photograph of the earth was taken en route to the moon during the final Apollo flight. The continent of Africa and the island of Madagascar are clearly visible. Egypt, the Red Sea and the Arabian Peninsula appear at the top. The field of snow and ice at the bottom is Antarctica.

Mars *(top)* and the moon, shown in proportion to their true sizes relative to the earth. Because of the moon's small size, its interior cooled rapidly and its volcanoes became extinct three billion years ago. Mars is larger and retained its volcanoes for a longer time, and also probably its water and atmospheric gases (pp. 134-143). The red spot at upper right on the Mars photograph is an extinct volcano.

of the ancient lunar highlands dating back to the beginning of the solar system. The electrically powered Rover is parked to the right of the landing craft.

Mare Imbrium, the left eye in the face of the man-in-the-moon, was formed four billion years ago by the greatest meteorite collision in the moon's history. Several hundred million years after the collision, the moon's interior partly melted and lava filled the gigantic impact crater to form the surface of the "sea."

THE SURFACE OF MARS. A scene on the plain of Chryse photographed on the surface of Mars by Viking. A part of the lander appears at the lower right. The orange-red color of the soil is believed to be caused by iron oxide, the same compound that colors the deserts of the Southwest. The salmon-yellow color of the sky near the horizon results from particles of soil stirred into the atmosphere by frequent Martian dust storms. Away from the horizon the sky changes to its normal deep blue color.

MOUNT OLYMPUS. This extinct Martian volcano is larger than the largest volcanoes on the earth. The summit, projecting above the clouds, is 70,000 feet high. The large crater at the summit is caused by the collapse of rock into pools of lava below. The smaller craters on the flanks are probably metorite impacts.

JUPITER. This photograph of Jupiter was taken by cameras on the Voyager spacecraft at a distance of 17 million miles. The bands circling the planet are belts of moving clouds made of crystals of frozen ammonia. The temperature at cloud-top level is minus 300 degrees. A zone of water clouds and comfortably warm temperatures lies beneath the ammonia cloud cover. The Great Red Spot *(lower left)* and several smaller white spots are mammoth storm centers.

THE FOUR LARGEST MOONS OF JUPITER. This montage shows Jupiter *(background)* and its four largest moons — Io, Europa, Ganymede and Callisto — seen from a vantage point near Callisto *(lower right)*. Callisto and Ganymede *(lower left)* are large drops of liquid water with frozen crusts of rock and ice. Io *(upper left)* and Europa *(center)* are composed largely of rocky materials like our moon.

THE RED SPOT. This detailed view of the Red Spot, taken at a distance of three million miles from Jupiter, reveals patterns of violent turbulence in the atmosphere of the Giant Planet. The Red Spot, 15,000 miles in diameter, is believed to be a gigantic anchored hurricane. It has persisted on Jupiter since the planet was first observed through a telescope by Galileo 300 years ago. The origin of the red coloring is unknown.

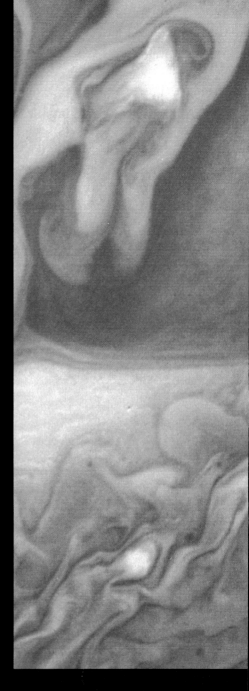

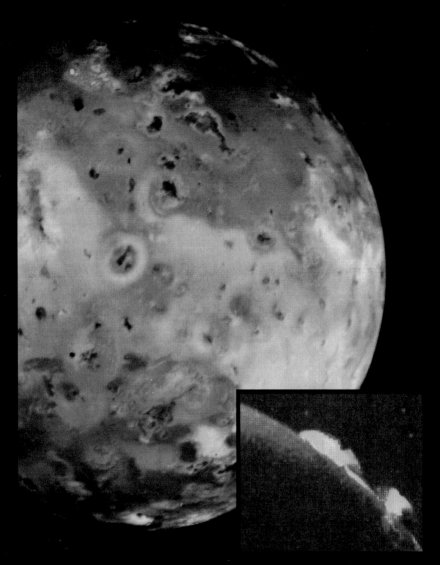

10. Io and the earth are the only volcanically active bodies in the solar system. The *inset* shows a volcanic eruption on Io, photographed by Voyager cameras.

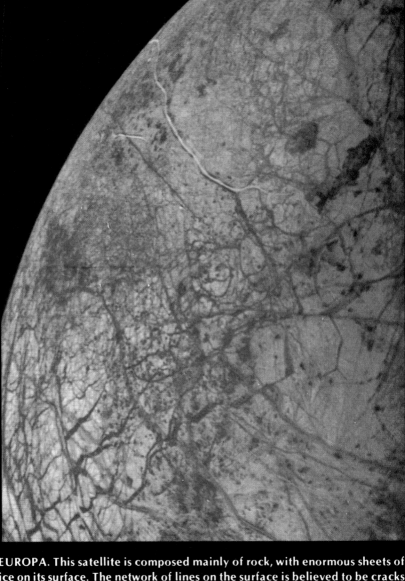

EUROPA. This satellite is composed mainly of rock, with enormous sheets of ice on its surface. The network of lines on the surface is believed to be cracks in the icy crust, perhaps similar to cracks in the earth's crust like the San Andreas fault.

SATURN AND TITAN. In August 1979 the Pioneer spacecraft reached Saturn after a six-year journey across a billion miles of space. A photograph of the planet (*top*), taken as the spacecraft approached, shows the famous rings, composed of innumerable tiny fragments circling Saturn like miniature moons. Details of the structure of the rings appear in a closer view (*middle*) from 585,000 miles. Pioneer also yielded a glimpse of Titan (*below*), a moon as large as the planet Mercury, and the only moon in the solar system known to have an atmosphere. Titan's atmosphere is rich in gases such as methane which can generate the building blocks of living matter. It also has a surface and may have a temperature high enough to support life.

because they appear to have substantial amounts of ice and water. Ganymede and Callisto are made largely of water; each can be described as a huge drop of liquid water with a core of rock or mud, and a thin crust of ice mixed with rock. Europa has less water or ice; it consists mainly of rock, with ice making up perhaps 5 to 10 percent of its bulk. Much of Europa's ice is lying on its surface; the Voyager photographs of Europa reveal large, bright areas, which are enormous ice sheets covering a substantial part of the planet. The photographs also show marks on the surface which may be fractures or faults, like the San Andreas fault in California. The cracks indicate that Europa is also volcanically active, or has been so in the past.

The composition of the icy satellites supports current scientific theories on the origin of the solar system.

Jupiter and its moons—some of them as big as planets—can be thought of as a small solar system inside a larger one. We would expect Jupiter's nearer moons, or "planets", to have practically no water, because the heat released when Jupiter formed would have driven off water and other volatile substances. This reasoning fits in with the indications that Amalthea and Io, the innermost satellites are made of nearly pure rock, with little water.

Europa, somewhat farther from Jupiter's early heat, would have retained a little more of its primordial water, according to the theory. The ice sheets covering Europa support this conclusion also. Ganymede and Callisto, roughly 500,000 and 1,000,000 miles from Jupiter, respectively, would have felt their parent planet's heat only weakly, and would have retained nearly all their primordial water, again in agreement with the theory.

Our solar system displays similar properties. The inner planets are composed mainly of rock and iron, with only small amounts of water and other volatiles; the earth, for example, contains about one percent water. According to the theory, this is because the intense heat of the newborn sun drove the light gases and volatiles from the inner regions of the solar system. Farther out, at the distance of Jupiter or Saturn, the sun's heat was weakened by distance, and the cloud out of which the planets formed retained its light gases—

including hydrogen and helium—as well as volatile substances, such as water. That is why Jupiter is made largely of hydrogen and helium. The moons of Jupiter lack hydrogen and helium because the gravity on these small bodies was too weak to hold the light gases; however, they were able to retain the water.

Because of the parallels between the formation of our planetary system and the formation of Jupiter and its moons, the exploration of Jupiter tests theories on the origin of the planets, which is one of the great riddles of Genesis. Jupiter seems likely to illuminate another mystery: How did life arise on the earth? Although Jupiter is a strange, unearthly body, and hostile to most terrestrial organisms, it has a special property that may lead to an understanding of that great scientific puzzle.

The atmosphere of Jupiter contains abundant amounts of the gases ammonia and methane,* in addition to hydrogen and water. Laboratory experiments have shown that these four gases— ammonia, methane, hydrogen and water—when mixed together and energized by an electric spark, produce copious amounts of amino acids—the molecular building blocks of living matter. This fact has given rise to a theory of the origin of life which is held in high regard by most scientists. The theory is discussed in more detail in Chapter 12. In brief, it states that the earth's atmosphere was composed of substantial amounts of methane and ammonia 4.6 billion years ago, when our planet was newly formed. Strokes of lightning—represented by the electric sparks in the laboratory experiment—energized the gases of the earth's atmosphere and produced a rich yield of amino acids. The amino acids, accumulating in the oceans, built up a nutritional broth. Random collisions in the broth, occurring again and again over millions of years, linked small molecules into long ones, and finally produced a molecule on the threshold of life.

*On a planet as large as Jupiter, the force of gravity is so great that most of the gases of the planet's original atmosphere will remain with it throughout its lifetime. Not even the lightest gases, hydrogen and helium, can escape. The atmosphere of Jupiter contains these gases in abundance, and it also contains the gases that are common compounds of hydrogen, such as ammonia, methane and water vapor.

Once the threshold of life was crossed, evolution commenced and the laws of natural selection came into play to produce the variety of plants and animals that now exist on our planet.

The life-giving gases—ammonia, methane and hydrogen—have long since vanished from the earth, except for trace amounts released from barnyards and rotting vegetation. (Pockets of methane—known as natural gas or coal gas—are still present underground.) The reason for their disappearance is that they contain atoms of hydrogen. Ultraviolet rays from the sun break up the molecules of methane and ammonia and release the hydrogen atoms, which then escape from the earth because the pull of its gravity is too weak to hold them. Jupiter's gravity, however, is so strong that not even hydrogen atoms can escape. This is the explanation for the presence of hydrogen compounds, such as methane and ammonia, on Jupiter.

Science has never been able to check its theory of the origin of life by searching the fossil record, because the earliest forms of life, lying on the threshold between the living and the nonliving worlds, were very fragile, and all traces of them have disappeared. And of course, no molecules on the threshold of life are being formed today, because the essential gases—methane and ammonia—also disappeared a long time ago.

Perhaps the theory can be checked in the laboratory, by putting together amino acids and other molecular building blocks and trying to create life out of nonliving matter. This prospect fires the imagination of the scientist. He looks into his flask and sees the lightning of a storm in the young earth's atmosphere; he smells the pungent aroma of ammonia; amino acids are forming; soon the first living organisms will emerge. . . .

But they never do. The elements of living matter accumulate in the flask, but no life climbs out. Why does the experiment fail? Because it lacks a key ingredient. The missing ingredient is time—millions on millions of years of ceaseless experimentation.

But on Jupiter, which still has its ammonia and methane, experiments on the origin of life have been going on for four billion years, and are still continuing. Even the electric sparks are present;

the Voyager photographs showed a brightening in some of the stormiest regions of the Jupiter atmosphere that suggested lightning discharges on a scale dwarfing terrestrial experience. With all the conditions of the laboratory experiments duplicated in the atmosphere of Jupiter, it seems likely that the building blocks of life may exist there in great numbers. Many interesting products of their collisions may also be present. It is the crucible of life.

Could life on Jupiter have progressed past the threshold and evolved to higher levels? Any such life would be small, squat and toadlike, adapted to withstand the crushing force of Jupiter's gravity—yet, perhaps, with a highly developed brain. This prospect seems less likely. On the earth, evolution proceeded rapidly in its early stages because the amino acids and other building blocks of life became concentrated in the shallow seas of our planet, and created a fairly thick soup in which collisions occurred frequently. These collisions, which linked small molecules together to form large ones, were the first step in the evolution of life. But Jupiter has no seas to catch the amino acids: They remain dispersed throughout the atmosphere, colliding infrequently, and the pace of evolution is correspondingly reduced. We expect that evolution might reach the threshold of life on the giant planet, and possibly cross it, but not that it would progress far beyond that point.

Still, Jupiter affords a unique opportunity to find out how life arose on the Earth, and how it may arise elsewhere in the cosmos.

A spacecraft called Galileo is scheduled for launch to Jupiter in 1982. Galileo will reach the planet in 1985, and drop into an orbit around it, from which it will eject a probe designed to fall toward Jupiter, radioing back information on the conditions that it encounters as it descends through the atmosphere. The probe will carry a mass spectrometer, an instrument that measures the masses of atoms and molecules accurately, which may provide the first direct information on the presence of the large molecules that could be building blocks of living matter. This instrument should be able to get down at least to the level of the water clouds on Jupiter, where the density of the atmosphere is high, collisions are more frequent, the temperature is comfortable, and conditions for the onset of chemical evolution are fairly favorable.

Galileo will bring us to the threshold of a serious search for life in the outer reaches of the solar system. A plan on the drawing boards calls for the subsequent exploration of Titan, largest moon of Saturn, which is important to that search because, unlike any other moon, it possesses its own atmosphere. Titan, like Jupiter, is also rich in methane and other gases helpful to the evolution of life. It may be even more favorable than Jupiter for the onset of evolution because it is likely to have a solid surface on which pools of liquid can accumulate for more rapid progress in nature's experiments. If the search for Martian life fails, Titan will offer the next opportunity to assess the prospects for life in the cosmos.

12 THE AGE OF THE EARTH

In 1648 James Ussher, Archbishop of Armagh, pronounced the creation of the earth to have occurred in 4004 B.C. A span of roughly 6000 years for the age of the earth, based on the biblical genealogy, was accepted by nearly everyone until the beginning of the nineteenth century, when geologists and naturalists began to suspect that the earth must have existed for a much longer interval of time.

During the course of the nineteenth century, geologists observed that land is stripped from the continents by water erosion; streams and rivers wear away their banks, carrying a small amount of soil from the highlands to the seas every year. Although only a minute fraction of the mass of the land is removed each year in this way, apparently the process has been going on for a very long time, for there are places in which one can see that an entire mountain range has been worn down to its roots. Usually it is difficult to detect these changes, but in a few special locations their effect can be seen very clearly. The Grand Canyon is the most striking example. There the Colorado River has cut through the earth's crust like a scalpel through tissue, exposing a clean record of the past. The history of that part of the earth is laid bare for everyone to see.

The cliffs of the Grand Canyon show that on no less than three successive occasions mountains have been raised to great heights by the heaving and buckling of the earth's crust, eroded away by running water, and created again in later upheavals. In other places, such as the Sheep Mountain in the Big Horn basin in Wyoming, the roots of a single mountain lie clearly exposed. The signs of erosion can be seen all around us on the face of the earth. Running water is the most effective agent, but glaciers and the sand-blasting action of dust-laden winds also take their toll. The sense of never-ending change on the earth came most strongly to James Hutton, the father of modern geology, who wrote in 1795:

> . . . from the top of the mountain to the shore of the sea . . . everything is in a state of change; the rock and solid strata slowly dissolving, breaking and decomposing, . . . the soil traveling along the surface of the earth on its way to the shore; and the shore itself wearing and wasting by the agitation of the sea.

How long has the surface of the earth been wasting away in this fashion? We can measure the amount of material which is removed each year from the American continent, for example, by the major rivers such as the Mississippi. It is estimated that 800 million tons of soil are washed away to the sea from the continental United States in a year. At this rate the level of the land is lowered by one foot in 10,000 years. A plateau two miles high is removed in 100 million years. Because this thickness of material has been worn away and replaced several times in some parts of the globe, it follows that the earth must be at least several hundred million years old. Moreover, the ceaseless cycle of erosion and uplifting has been going on as far back into history, and as deep into the crust of the earth as we are able to see. There is no clue as to where and when to stop. For all the geologist can tell, the earth may have existed forever. Hutton said: "We find no vestige of a beginning."

Several generations later, Charles Darwin undertook a study of living creatures and their relation to the fossils of ancient animals preserved in the rocks. There are now approximately one million species of animals on the face of the earth. Darwin saw that the forms of life existing on the earth today have evolved gradually out of earlier and simpler beginnings through a succession of very small changes. From the examination of a large number of fossil skeletons, it can be seen that over a period of 60 million years the modern horse, for example, has evolved from a small, five-toed animal the size of a fox terrier, as a result of a long series of minor modifications. Darwin noted that in modern animals the changes from one generation to the next are imperceptibly slight, too slight to be detectable within the

THE GRAND CANYON: AN INCISION IN THE CRUST OF THE EARTH. The sides of the Grand Canyon show the remains of three mountain ranges which have been raised up and successively worn away to their roots by the erosive action of running water. The Colorado River *(below, running from left center to bottom right)* has cut through the crust of the earth to a depth of 6000 feet to expose the record of these changes. It reveals layers of the earth's crust which were formed as long ago as one billion years.

lifetime of one person, or even within the memory of the human race. He concluded that a substantial change in the form of an animal must have required thousands if not millions of generations, and that a vast amount of time must have elapsed from the beginning of the fossil record to the present. Furthermore, he suggested that a long interval must have elapsed prior to the appearance of fossils, in which soft-bodied animals filled the primitive seas without leaving any trace of their existence.

Exactly how long have these changes in the forms of living creatures been going on? Darwin could not answer this question, but he, too, felt intuitively that the earth must have existed for many hundreds of millions of years.

Darwin's views regarding the antiquity of the earth clashed with the opinions of Lord Kelvin, the British mathematical physicist. Geophysical measurements had revealed that a small amount of heat steadily flows from the interior of the earth to its surface. According to Kelvin, this flow of heat resulted from the fact that the earth was a molten mass when it was newly formed. During the course of millions of years, as it lost heat from the surface, it gradually cooled and hardened. Kelvin calculated how long it would take for the earth to cool to its present temperature, assuming that heat had been flowing out of the interior of the planet throughout its past history at the same rate at which it flows out today. The calculations indicated that the earth had been cooling down for about 40 million years. Kelvin announced that this was the age of the earth. He thought that 40 million years was, if anything, a generous estimate, because when the earth was a younger and hotter planet, it probably lost heat at the surface more rapidly than it does today.

But, according to Darwin, a far greater time than 40 million years was needed for species of plant and animal life to have developed by the slow process of natural selection. If Kelvin's calculations were correct, Darwin's theory of evolution must be wrong.

Darwin was very upset by this development. In 1869 he wrote: ". . . I am greatly troubled at the short duration of the world according to Sir William Thomson [Lord Kelvin] for I require for theoretical views a very long period before the Cambrian formation." The strain must have been grievous, for he seems to have taken a personal dislike to Kelvin, referring to him, in a letter to Alfred Wallace, as "the odious spectre." But Kelvin was confident; in 1873 he said: "We find at every turn something to show . . . the utter futility of [Darwin's] philosophy." By 1893 he had reduced his estimate of the age of the earth to 24 million years, squeezing Darwin relentlessly.

In the conflict between Darwin's intuition and Kelvin's mathematical physics, intuition triumphed, for Kelvin had omitted a major factor from his calculations in the 1870s. The missing factor did not come to light until 1904, twenty-five years after Darwin's death. In that year Rutherford discovered that radioactivity releases appreciable amounts of heat. The earth contains radioactive substances—thorium, uranium and potassium—in its interior. Kelvin had not known of the existence of these substances. According to Rutherford's measurements, they released enough heat to extend the time of cooling of the earth far beyond Kelvin's estimates.

Rutherford discussed the implications of his experiment in a lecture given in 1904, which Kelvin attended. Afterward Rutherford said:

"I came into the room, which was half dark, and presently spotted Lord Kelvin in the audience and realized that I was in for trouble at the last part of my speech dealing with the age of the earth, where my views conflicted with his. To my relief, Kelvin fell asleep, but as I came to the important point, I saw the old bird sit up, open an eye and cock a baleful glance at me! Then a sudden inspiration came, and I said Lord Kelvin had limited the age of the earth, provided no new source was discovered. That prophetic utterance refers to what we are now considering tonight, radium! Behold! the old boy beamed upon me."

The discovery of radioactivity released the naturalists from the strait-jacket of Kelvin's calculations and gave them as much time as they needed. It is curious that the same discovery also yielded the first indication of the actual age of the earth. The method of measurement was conceived, once again, by Lord Rutherford, and worked out in detail by B. B. Boltwood of Yale University. Their technique was based on a discovery made by Marie and Pierre Curie. All rocks

CHARLES DARWIN (1809-1882): DEFENDER OF AN ANCIENT EARTH. This photograph was taken in 1854, when Darwin was forty-five years old. At that time he had been thinking about fossils and their relation to living animals for nearly twenty years. He had concluded that the forms of life now on the earth have developed out of vanished species by a succession of innumerable small changes. These changes, imperceptible from one generation to the next, must have occurred over an exceedingly long period of time. Darwin was convinced that the earth was a very old planet.

LORD KELVIN (1824-1907): EXPONENT OF A YOUNG PLANET. Lord Kelvin, one of Britain's greatest physicists, disputed Darwin's views on the age of the earth. Kelvin's calculations showed that the earth had been a molten, uninhabitable body of rock no more than 40 million years ago. This interval of time was too short for life to have evolved to its present variety and complexity by a succession of small changes, as Darwin's theory of evolution proposed. Darwin died in 1882, deeply troubled by Kelvin's criticism. In 1907, twenty-five years after Darwin's death, Rutherford discovered that radioactive substances buried in the earth's interior released sufficient heat to invalidate Kelvin's calculations.

carry a very small amount of radioactive uranium. The Curies had shown that atoms of radioactive uranium undergo a natural transformation to the element lead. The ratio of lead to uranium, Rutherford reasoned, will reveal the length of time that the rock has existed. As the atoms of uranium decay one by one, the concentration of lead in the rock must steadily increase. The amount of lead present, Rutherford reasoned, will reveal the length of time over which the decay from uranium to lead has been occurring. If the rock is newly formed, few of its uranium atoms will have decayed into lead, and the amount of lead will be small; if the rock is old, many of its uranium atoms will have decayed into lead, and the amount of lead will be large.

Laboratory measurements show that 50 percent of the uranium in a rock will change to lead in the course of 4.6 billion years. This is known as the half-life of uranium. Boltwood found a large amount of lead in the rocks he studied, indicating that they had existed for an appreciable fraction of the uranium half-life, perhaps for as long as a billion years. In subsequent years still older rocks were dated in this way, yielding ages ranging up to 3.3 billion years.

Meteorites, which are pieces of extraterrestrial rock that occasionally collide with the earth, have also been studied by the same technique. The results show that most meteorites are about 4.5 billion years old. Meteorites have not been worked over by erosion, and subjected to a variety of chemical and physical changes, as in the case for the rocks on the surface of the earth. For this reason, it is believed that they give a better indication of the original state of matter in the solar system than is given by terrestrial rocks. Therefore, 4.6 billion years, the age of meteorites, is assumed to be the age of the solar system and the age of the earth.

13 THE DAWN OF LIFE

The earth began its existence 4.6 billion years ago, circling around the newborn sun. Formed out of inert atoms of gas and grains of dust, our planet was surely a sterile body of rock at the beginning. The waters of the primitive oceans were devoid of life; their waves lapped at barren shores, uncarpeted by vegetation. Yet today plants grow everywhere; the continents crawl with a million varieties of animal life; 20,000 kinds of fishes inhabit the seas. How and when did this rich variety of living forms appear on our planet?

Advances in science in recent decades have uncovered facts about the nature of living organisms which lead for the first time to a scientific explanation of the origin of life. It now appears likely that the first living creatures on the earth evolved spontaneously out of the inert chemicals that filled the atmosphere and oceans of the planet in the early years of its existence. Three discoveries lead to this conclusion.

First, biologists have shown that all living organisms on the face of the earth depend on two kinds of molecules—amino acids and nucleotides, which are the basic building blocks of life—just as the physicists have shown that all matter in the universe is constructed out of three building blocks—the neutron, the proton and the electron.

Second, chemists have manufactured these molecular building blocks of life in the laboratory out of simple chemicals, under conditions resembling those that existed on the earth when it was a young planet.

Third, an object has been discovered which links the nuclei, atoms and molecules of the physical universe to the complex organisms of the living world. This object, called the virus, lies on the borderline between inanimate matter and life. Its existence gives credibility to the notion that life evolved out of nonliving chemicals.

The basic building blocks of life are more complicated than the building blocks of the physical world. Twenty different kinds of amino acids play a critical role in living creatures, and five different kinds of nucleotides.*

Furthermore, each amino acid or nucleotide is, itself, a rather complex molecule made up of approximately thirty atoms of hydrogen, nitrogen, oxygen and carbon, bound together by electrical forces of attraction.

The amino acid and the nucleotide have very different functions in the chemistry of life. Within the cell the amino acids are linked together into very large molecules called proteins. One class of proteins, called structural proteins, make up the structural elements of the living organism—the walls of the cell, hair, muscles and bone. The structural proteins are like the steel framework and walls of a building. The other type of protein is called the enzyme. Many kinds of enzymes exist; each kind controls one of the many chemical reactions that are necessary to sustain the life of the organism.

All proteins in all forms of life, plant and animal, are constructed out of the same basic set of twenty amino acids. One protein differs from another only in the way in which its constituent amino acids are linked together. However, these differences are all-important. The distinction between a man and a mouse, in both appearance and personality, depends entirely on the differences between the proteins contained in the cells of their bodies.

Proteins are assembled within living organisms by the second set of building blocks—the nucleotides. Nucleotides are joined together within the cell to form very long chains, called nucleic acids. The most important type of nucleic is called deoxyribonucleic acid, or

*It is more correct to say that there are five *nucleotide bases.*

DNA for short.* DNA is the largest molecule known, containing, in advanced organisms such as man, as many as 10 billion separate atoms. The size of the DNA molecule is understandable when we consider the complexity and importance of its functions in the living cell. The DNA molecule is the most important molecule in every living organism, even more important than the protein, because it determines *which* proteins will be assembled; the DNA molecule has the master plan for the organism.

How does DNA control the assembly of proteins in the cell? The general features of the process began to emerge during the decade of the 1950s, although many of the details are still not clearly understood. It appears that the separate amino acids and nucleotides float freely in the fluid of the cell. The DNA molecules which direct the assembly of proteins are located in the center of the cell. In the first step, free nucleotides are attracted to a segment of one of the DNA molecules at the center. They line up alongside the DNA segment to form a replica of it. In the second step, the replica detaches itself from the master DNA chain, and drifts off into the cell; it is a messenger which carries instructions from the DNA into the body of the cell for the assembly of one particular kind of protein. In the third step, another molecule enters the picture. This molecule serves as a connecting link, bringing the amino acids in the fluid of the cell to the appropriate places alongside the messenger. There are twenty kinds of connecting links, one for each kind of amino acid. Each of the connecting links attracts one and only one of the 20 amino acids. When, in the course of chance collisions, the right kind of amino acid comes into contact with the end of the connecting link designed for that particular amino acid, it is held fast there. At the other end of the connecting link is another set of molecules, making up a surface of nooks and crannies so constructed that it can fit only into the appropriate place along the length of the messenger. When the connecting link takes its place along the messenger, it adds the amino acid to the

---

*Only four of the five important nucleotides enter into the structure of DNA. The fifth nucleotide belongs to another type of nucleic acid.

chain of amino acids that has already been built up. When a chain of amino acids has been assembled along the full length of the messenger, the assembly of amino acids into a protein is complete. The assembled chain then detaches itself from the messenger and drifts off into the fluid of the cell.

By this rather complicated process, the essential proteins are built up within an animal in accordance with the order of the nucleotides in its DNA molecules. The segments of the DNA molecule are "read" like the words of a book. Each DNA segment, controlling the assembly of one protein, is a word; each nucleotide within a segment is a letter; the order of the letters provides the meaning of the word—that is, the protein to be assembled. The full set of DNA molecules in the cells of a mouse direct the assembly of its amino acids into mouse proteins.

How is the plan for the assembly of the right kind of proteins passed from one generation to the next? How do progeny inherit their characteristics from their parents? The answer lies in a most extraordinary property of the DNA molecule—the ability to make a copy of itself. The mechanism by which DNA copies itself was discovered in 1953 by an Anglo-American team, James D. Watson of Harvard University and Frances Crick of Cambridge University. This discovery is one of the most important single scientific events of the twentieth century to date. I have described the DNA molecule as a chain of nucleotides; but Watson and Crick found it is not a *single* chain of nucleotides; it consists, rather of *two* chains, joined together at regular intervals by molecules which run between them like rungs of a ladder. At the middle of each rung of the ladder is a weak spot, which is easily broken. During the early history of a cell the two strands remain connected, but when the cell has attained its full growth, and the division into two daughter cells is about to commence, the weak connections running down the middle of the ladder break, and the double strand separates into two single strands. Each of the single strands then collects unto itself new nucleotides out of the pool of nucleotides floating in the cell, and assembles them into a new double-stranded molecule. There now exist two identical DNA molecules, where formerly there was one. The two DNA molecules separate, and move to opposite corners of the cell; the cell then divides into two daughter cells, each containing one complete set of DNA molecules. Thus, each of the daughter cells contains a copy of the volume of genetic information which had been in the parent cell. This is the way in which the shape and the character of a plant or an animal are transmitted from generation to generation.

In summary, the DNA molecule controls the assembly of proteins, and the proteins determine the nature of the organism. Each living organism has it own special set of DNA molecules; no two organisms have the same set unless they are identical twins. However, *the basic nucleotides and amino acids are the same in every living creature on the face of the earth, whether bacterium, mollusc or man.*

With this fundamental property of living creatures in mind, one can appreciate the importance of a critical experiment performed in 1952 by Stanley Miller. Miller, now a professor of biochemistry at the University of California, was then a graduate student working on his Ph.D. thesis under Harold Urey. At the suggestion of Urey, Miller mixed together the gases—ammonia, methane, water vapor and hydrogen—which were abundant in the parent cloud of the earth, and which were probably abundant in the earth's primitive atmosphere. He circulated the mixture through an electric discharge. At the end of a week, Miller found that the water contained several types of amino acids. Subsequently, nucleotides were created in the laboratory under similar conditions. In similar experiments amino acids and nucleotides have been manufactured out of a variety of gas mixtures, using various sources of energy—bombardment by alpha particles, irradiation with ultraviolet light, and simple heating of the ingredients. The results of all these experiments, taken together, demonstrate that the molecular building blocks of life could have been created in any one of many different ways during the early history of the earth.

Amino acids and nucleotides might have been formed on the earth in this way 4.6 billion years ago, by the discharge of lightning in primitive thunderstorms, or by the action of solar ultraviolet rays from the sun. We can guess what happened thereafter. Gradually the critical molecules drained out of the atmosphere into the oceans, building up a nutrient broth of continuously increasing strength. Over a long period of time the concentration of amino acids and nucleotides increased, until eventually a chance combination of building blocks produced still more complex molecules—primitive proteins and nucleic acids. With the further passage of time, cells developed; many-celled organisms appeared; and living organisms were started on the long road to the complexity of the creatures which exist today.

This is a nice story that has emerged from the union of astronomy, biology and chemistry. Yet, it is hard to believe that existing forms of life, in all their variety and sophistication, can be traced back to simple chemicals. Is there any direct evidence for the development of life out of nonliving molecules?

The answer is: Yes, there is an entity, very common in the world today, which possesses, at the same time, the attributes of a nonliving molecule and the attributes of a living organism. This entity is the virus—the smallest and simplest object which can be said to be alive.

The existence of viruses first came to light at the end of the nineteenth century, in the course of a series of experiments designed to reveal the cause of a disease affecting tobacco plants. It was found that the juice pressed from the leaves of infected plants could transmit the infection to other plants. Apparently, the infection was transmitted by bacteria contained in the fluid. But when pressed through a fine filter, which screened out all visible bacteria, the fluid still retained its power of infection. In 1898 a Dutch botanist, Beijerinck, suggested that the disease was not caused by a germ, but by a poisonous chemical. Beijerinck called the chemical a "virus," which is the Latin word for poison.

Further research revealed that viruses are the cause of many diseases, including smallpox, influenza, infantile paralysis and the common cold. Medical and biological interest in viruses intensified during the first decades of the twentieth century. Gradually the suspicion developed that the virus was no ordinary chemical. A variety of experiments suggested that the virus, although too small to be seen under the microscope, possessed the basic attribute of living organisms—the ability to reproduce itself.

Still, the evidence for the living virus was indirect; no one had yet seen one in the act of reproduction. But in the years after the Second World War a new instrument was perfected, which provided the biologists with a powerful tool for the study of small organisms. This instrument was the electron microscope. Ordinary microscopes, in which the object under study is illuminated by rays of light, are limited to a magnifying power of approximately 2000. The smallest bacteria, which are a hundred-thousandth of an inch in size, can just barely be seen in these microscopes. But the electron microscope, which direct a beam of electrons at the object instead of a beam of light, can produce magnifications as high as several hundred thousand diameters. It is possible to photograph a single protein molecule with these instruments; if a further improvement in magnification by a factor of 50 can be achieved, the electron microscope will be able to photograph individual amino acids and nucleotides.

Under the electron microscope the virus finally became visible, and all the important details of its structure were revealed. It was found that viruses come in many shapes—round, cylindrical, polyhedral and with tails. They also come in many sizes. The largest is as large as a small bacterium; the smallest, which is a millionth of an inch in diameter, is smaller than many nonliving molecules. Viruses bridge the gap in size between the inanimate and animate worlds.

Yet these tiny particles are indisputably alive. Chemical studies show that they contain DNA—the molecular blueprint of life and the means by which every living creature reproduces itself. They also contain a substantial amount of protein, in the form of a protective coat wrapped around the precious, delicate strands of DNA. But they contain very little else. In particular, they have none of the sugar and fat molecules that provide energy for the chemical reactions in other living creatures. They also lack free nucleotides and amino acids, out of which all other organisms make proteins and assemble copies of themselves.

How, then, without a source of energy, and without the materials essential for growth and reproduction, do viruses live?

The answer is clearly revealed by the electron microscope. A virus, by itself, is not alive. If a solution of virus particles is carefully dried out the viruses will stick together in a symmetric pattern to form a crystal as geometric—and as lifeless—as a crystal of salt or a diamond; left undisturbed, the crystal remains inert for years. But, dissolved again in water, and placed in contact with living cells, the molecules of the crystal spring to life; they fasten to the walls of the cell, dissolve a small opening in the cell wall, and, through the opening, inject their DNA into the cell. Once inside the cell, the virus DNA seizes control, displacing the original DNA of the cell and extablishing itself as the master of all futher chemical activity. The full molecular resources of the invaded cell—the energy-giving fats and sugars, the amino acids and the nucleotides—are commandeered and employed in the assembly, not of the proteins needed by the invaded cell, but of the proteins needed by the virus. At the same time, the virus gathers the free nucleotides floating in the cell fluid and assembles them, not into copies of the DNA of the invaded cell, but into copies of itself. The virus even secretes an enzyme which breaks down the existing DNA within the cell into its component nucleotides, in order to have more of these precious units available for making replicas of itself.

When several hundred protein coats and virus DNAs have been assembled, the cell is milked dry. The coats wrap around the virus DNA molecules to form complete viruses, while the original virus, in a final step, secretes an additional enzyme which dissolves the cell walls. An army of virus particles marches forth, each seeking new cells to invade, leaving behind the empty, broken husk of what had been, an hour before, a healthy living cell. The operation is simple, ruthless and effective. It is executed by an organism which is, in the smallest viruses, only 200 atoms wide. The virus is truly the link between life and nonlife—the bridge between living and non-living matter.

THE ORIGIN
OF LIFE

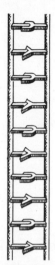

THE STRUCTURE OF DNA. Watson and Crick discovered the structure of the DNA molecule in 1952. DNA resembles a ladder *(above, left)*; each rung of the ladder is a linked pair of nucleotides, represented by the symbols

Actually, the ladder is twisted into a double spiral, called a double helix *(top, right)*.

The model at right shows the manner in which the double spiral of DNA is constructed out of individual atoms. Each plastic ball in the model represents one atom. This short segment of DNA contains approximately 1000 atoms; in the higher animals, the complete molecule may contain 10 billion atoms. In the photograph the author *(far right)* questions Dr. Gordon M. Tomkins, Chief of the Division of Molecular Biology in the National Institutes of Health, regarding the structure of DNA.

1

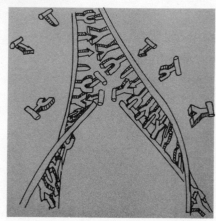

2

THE DNA MOLECULE COPIES ITSELF. When Watson and Crick discovered the "twisted ladder" construction of DNA, it became clear how the traits of an individual are passed from generation to generation.

The record of individual traits is stored in the DNA molecules which reside in the cells of each organism. In every organism more advanced than the lowly virus, growth and reproduction involve the division of these cells. When a cell is about to divide, its DNA molecules unzip, the "ladder" untwists, the rungs break at their midpoints, and the DNA begins to separate into two parallel strands (1).

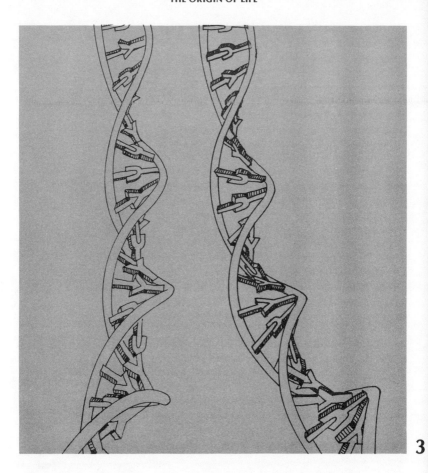

3

Throughout this process, unattached nucleotides float nearby in the fluid of the cell. In the next step, each strand of the unzipped DNA collects new nucleotides from the molecules surrounding it in the cell to form a new complete "ladder" (2).

The result is two "ladders," replicas of the original DNA molecule (3). Each replica becomes the center of a new daughter cell. In this way the master plan of the organism is passed from cell to cell and from generation to generation.

HOW THE MASTER PLAN IS CARRIED OUT.

How does DNA direct the assembly of amino acids into vital proteins? The DNA molecules reside in a nucleus at the center of the cell, enclosed by a membrane. When the DNA molecule is about to make a protein, a segment of the DNA — one gene — splits into two strands, and a series of nucleotides assembles along one of the strands (1), copying the master sequence.

TRANSFER RNA

The assembled string of nucleotides, which is slightly different chemically from DNA, is called RNA. As the RNA molecule assembles, it peels off and goes out into the cell through an opening (2) in the membrane surrounding the nucleus. This RNA molecule is called "messenger RNA," because it carries the message from the DNA molecule to the rest of the cell containing the instructions for building a protein.

The cell contains another kind of RNA molecule resembling a stick *(upper left)*, with attachments for amino acids on one end of the stick and nucleotides on the other end. This molecule is called "transfer RNA" because it picks up amino acids (3) and transfers them to the proper spot on the messenger RNA (4), where they add to the growing chain of amino acids that makes up the complete protein (5). After each transfer RNA deposits its amino acid, it goes off into the cell to (6) find another amino acid and do its work again.

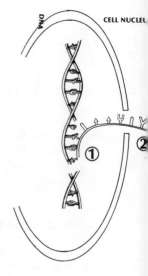

DNA

CELL NUCLEUS

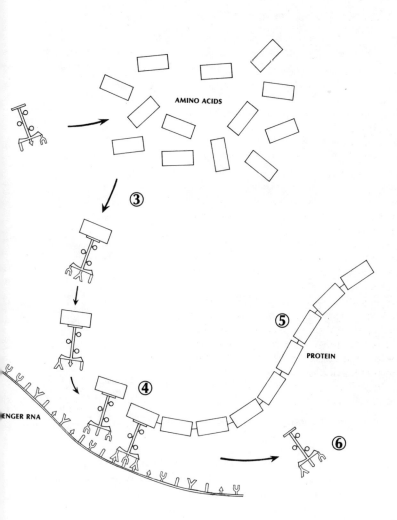

AMINO ACIDS

③

④

⑤

PROTEIN

⑥

MESSENGER RNA

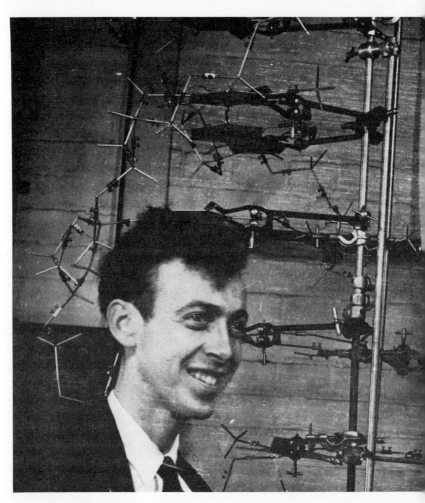

DISCOVERERS OF THE DOUBLE HELIX. The photograph *(above)* shows Watson *(left)* and Crick standing by their first successful model of DNA in the Cavendish Laboratory in Cambridge. Their discovery depended on results obtained by Rosalind Franklin *(opposite, right)* who took an x-ray photograph *(opposite, left)* of the DNA molecule so revealing of its properties that, Watson wrote later, "The moment I saw the picture my mouth fell open." A trained eye can see in this photograph the double-helix structure of DNA, the pitch of the helix, the spacing of the rungs, and the fact that the heavy part of the molecule — its backbone — is on the outside. Franklin's results also led Crick to what he later called "the crucial fact" that the two spirals making up the molecule do not run in the same direction like the stripes on a barber pole, but in opposite directions, making a twisted ladder.

A CRITICAL EXPERIMENT: THE BUILDING BLOCKS OF LIFE CREATED IN THE LABORATORY. Biochemists have manufactured the molecular building blocks of life in the laboratory out of simple ingredients. The first experiment of this kind was performed by Stanley Miller in 1952. He mixed together gases — ammonia, methane, water vapor and hydrogen — which were probably present in the earth's primitive atmosphere, and circulated them through a glass bowl containing an electric discharge. At the end of a week, Dr. Miller found that the water contained several varieties of amino acids. Subsequent experiments have created other molecular building blocks of life in the laboratory out of a variety of chemicals and under many different circumstances. Life may have developed on the earth out of such molecules, 3 or 4 billion years ago.

The drawing of the Miller apparatus *(below)* shows the round glass bowl in which amino acids were created when a spark passed between two electrodes. At right, Dr. Miller stands next to the original apparatus with which he performed his critical experiment.

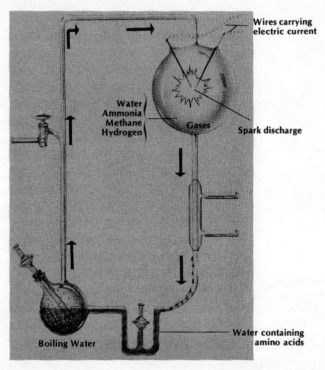

THE VIRUS: LINK BETWEEN LIFE AND NON-LIFE. The existence of the virus adds credibility to the notion that life evolved out of nonliving chemicals: the virus lies on the threshold between living matter and inanimate molecules; it is the simplest and smallest living particle, some viruses being only one-millionth of an inch in diameter.

Carefully dried out, a solution of virus particles forms completely inert crystals *(below)*. Dissolved in water and given access to living cells, the viruses composing the "crystals" come to life and attack their hosts *(opposite)*.

In this photograph, viruses have gathered around a sausage-shaped bacterium. Note the minuteness of the viruses relative to the bacterium. In the *upper right* corner are the remains of a bacterium which was attacked one hour earlier. Viruses entered this bacterium and consumed its chemicals in making replicas of themselves, leaving the dry husk of a previously healthy cell.

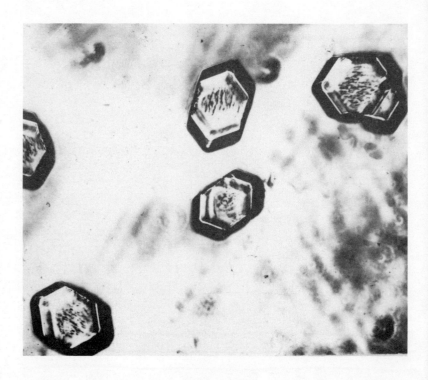

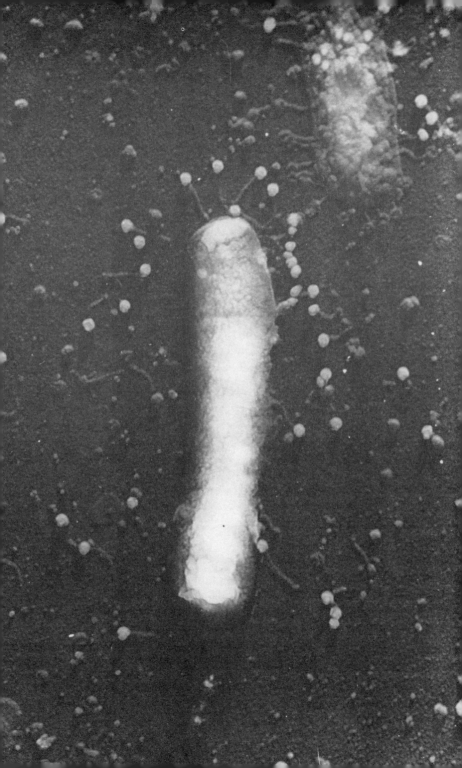

14 UFOs

The imagination of the scientist has seized on bits and pieces of research accumulated in many different fields of science, and fashioned out of them a picture of the origin of life on the earth. No living form existed on our planet in its infancy; the atmosphere was filled with a noxious mixture of ammonia, methane, water and hydrogen; peals of thunder rumbled across the sky; flashes of lightning occasionally illuminated the surface, but no eye perceived them; minute amounts of amino acids and nucleotides were formed in each flash, and gradually these critical molecules accumulated in the earth's oceans; collisions occurred between them now and then, linking small molecules into larger ones. During the course of a billion years the concentration of complex molecules increased; eventually a complete DNA chain appeared. Thus the threshold was crossed from inorganic matter to the living organism.

According to this story, life can appear spontaneously on any earthlike planet that offers a congenial climate. How many such planets are there? The astronomical evidence suggests that there are billions of planets in our galaxy, and billions more in other galaxies around us. In this great multitude of planets some will be too close to their parent stars, and too hot; others will be too far away, and too cold; a few will be just at the right distance to create the gentle warmth needed for the evolution of life. Let such favored planets be relatively few in number; let them be as rare as one in a million; no matter, *the number of planets suitable for life will still be 100,000 in our galaxy alone.*

How many of these favored planets actually bear life? No one knows; all may be barren. But the biological discoveries described in the last chapter suggest that this is not the case. Here is the evidence. All life is made from two dozen molecular building blocks; these building blocks of life are readily created in the laboratory under conditions that would prevail on any earthlike planet; the atoms that compose the building blocks of life are the same as the atoms that exist on every planet in the universe; and there is reason to believe that the same laws of physics and chemistry apply in every corner of the cosmos.

These facts give credibility to the idea of the UFO; it is an idea which is scientifically sound. Why should the chain of chemical reactions that led to life on the earth not occur on other planets, circling other stars? Why should the earth —an undistinguished speck of planetary matter —be the only body in the heavens to bear life? It is more reasonable to assume that man is not alone in the universe.

This reasoning does not constitute proof that UFOs are extraterrestrial. No concrete evidence exists for that belief. But if, in fact, some UFOs *have* originated beyond the earth, science has something to say about them. They cannot have come from another planet circling our sun, because no intelligent life exists in this solar system except on the earth. All the evidence regarding conditions on other planets in our solar system, described in the chapters on Venus, Mars and Jupiter points to that conclusion. It follows that the UFOs have come from another star.

There is the rub. The closest star to the sun is 25 trillion miles away, and it would take 100,000 years to cover that enormous distance with the fastest rockets known to man. We cannot make the journey; a trip to the stars is beyond our reach at the present time. Is it possible that others have succeeded where man would fail?

Considering this question, we must reflect that the universe is 20 billion years old, while the earth is roughly five billion years old. This means that some of the planets in the universe are 15 billion years older than the earth. Fifteen billion years is a long time in evolution; a mere billion years ago, the highest form of life on the earth was a mindless, worm-like creature. When we reflect on the scientific advances of the last 100 years, we realize that the advances which will occur in another billion years are beyond our imagination. Consider the history of man: we have existed as a human species for barely two million years; modern science is only 300 years old; and it is only a decade since we made the first modest journey to our nearest neighbor in space. The period in which our scientific knowledge has developed is an exceedingly narrow slice of time, sandwiched between the billions of years of evolution that preceded the emergence of man and the billions of years that lie before us in the lifetime of the solar system.

It is exceedingly unlikely that any society on another planet came into existence at the same moment of time, and developed at the same rate, so as to have arrived at precisely the same level of technology which we possess on the earth today. A difference of 100 years, which is the blink of an eye in the lifetime of a star or planet, has produced enormous changes in the scientific knowledge of our society. Some of these extraterrestrial societies, living on planets born later than the sun, must be primitive in comparison with us; others, with an earlier start, must have surpassed our achievements a long time ago.

It is to this latter group — the more advanced societies — that we should direct our attention, for we must expect that they will have mastered the techniques of space travel, with a greater skill than we can hope to achieve for many centuries. We must expect that these older, more advanced societies will reach us before we discover them. This conclusion lends further credibility to the idea of the extraterrestrial visitor.

Have such visits occurred? The first chapter of the Book of Ezekiel records a remarkable incident that took place several thousand years ago. After an account of what seems to be a landing and exploration by unusual beings, apparently metallic in construction, Verse 24 describes their departure: "And when they went, I heard the noise of their wings, like the noise of great waters. . . ." Anyone who saw a Saturn V rocket take off will remember that the thunderous roar sounded like Niagara Falls. Nothing manmade except the launch of a great rocket sounds like that.

When can we expect the next contact? Fairly soon. Until recently we were inconspicuous, but during the last 20 years we have made ourselves known throughout this part of the heavens. Since about 1960, television stations scattered across the earth have been spraying their signals into space at a million-watt level. In the course of those twenty years, the expanding shell of television signals has swept past about 40 stars like the sun. Old programs, moving away from the earth at the speed of light, have carried the message that intelligent life exists on this planet. These television programs make the earth the brightest radio star in our neighborhood of the Galaxy. If any of those 40 stars harbors intelligent beings, our presence is now known to them. As it took twenty years for our signals to reach those stars, it must take twenty years for their reply, traveling at the same speed, to get back. Unless man is alone in the cosmos, we can expect to receive a message by the end of this century.

15 MILLIONS OF GENERATIONS

Many planets revolve about other stars; there may be millions in our galaxy, and perhaps an infinite number in the universe. Doubtless, many are dead bodies of rock, washed by sterile seas. But on some, situated at favorable distances from their suns, the environment is suitable for the formation of nucleotides and amino acids — the building blocks of life. On these planets the chaotic succession of inanimate processes gives way to a pattern of chemical evolution, complex and self-reproducing.

Life appeared on the earth as the product of this sequence of events, at some point in the first billion years of its existence. The earliest organisms were very simple, scarcely more than giant molecules immersed in the primeval waters of the planet. During the billions of years that followed, those organic molecules developed into the rich variety of plants and animals that now live on the earth. What guided the course of evolution on this planet from the first primitive organisms to the complicated creatures of today? If life has arisen elsewhere, what guides the course of evolution on other planets? Is there a law in nature which controls the forms of life?

The fossil record contains clues to the resolution of this question. Thousands of skeletons and fossil remains mark the path by which life climbed upward from its crude beginnings. The initial steps along that path are not known; those first forms must have been fragile, for no trace of them remains. The earliest signs of life to appear in the record, already far advanced beyond the "living" molecule, are the deposits of simple one-celled plants called algae, and the shells of rod-shaped organisms resembling bacteria. These are found in rocks formed 3 billion years ago, when the earth was already more than a billion years old.

Very little happened for several billion years thereafter; at least, very little that has been preserved in the record of the rocks. But suddenly, 600 million years ago, the pace of evolution quickened. In rocks of that age the first hardbodied animals—corals, starfish, snails and trilobites—appear in great numbers. During the next 200 million years life exploded into a profusion of different forms. By 400 million B.C. all the major branches of the animal kingdom had developed.

At that time the highest forms of animal life were still confined to the waters of the planet, and the land was relatively barren. But 350 million years ago one class of aquatic animals—the fishes—developed air-breathing forms; some air-breathing fishes evolved into amphibians—the first vertebrate animals* to venture onto the land—and out of the amphibians, 50 million years later, came the reptiles. The reptiles were the first vertebrates to be completely emancipated from the water. Branches of the reptiles gave rise to the snake, lizard, turtle and bird; other branches produced the dinosaurs and their descendants, the crocodile and alligator; still other branches led to the mammals.

The dinosaurs ruled the earth for 100 million years, and during the long reign of these highly successful reptiles the mammals were kept in check and made little evolutionary progress. But suddenly, 70 million years ago, the dinosaurs disappeared. With their disappearance the mammalian stock flowered in a variety of forms, until, by 10 million B.C., the ancestors of most of the animals we see on the earth today, from aardvarks to zebras, had evolved. Two or three million years ago, late in this story, an animal recognizably similar to man appeared on the scene.

The record of these changes contains may gaps, but the segments which are present convey a clear message: Man has evolved slowly, during some billions of years, out of lower and simpler organisms. And, although the first part of the record is missing, it is probable

---

* A vertebrate is an animal possessing a backbone and an internal skeleton, as distinct from invertebrates such as the insect, for example, whose skeleton is external and surrounds its body.

that these lower organisms, in turn, came from nonliving molecules formed in the waters of the primitive earth.

In this book I have shown how the basic forces of nature—gravity, electromagnetism and the nuclear force—acting on the basic building blocks of matter, have lead, first to the synthesis of the elements within the interiors of stars; then, to the formation of the sun and planets out of those elements; and, finally, on the surface of one of these planets, to the formation of organic molecules lying on the threshold of life. I have shown how that threshold may have been crossed in the early years of the earth's existence. Throughout this long history my viewpoint has been that of the physicist, seeking to understand the essence of the world around him in terms of a few simple principles. One might call them the laws of physics. These laws are the distillation of all the observations regarding the physical world which have been acquired in thousands of years of human experience.

Now we come to the explanation of the subsequent course of events in the history of life, leading from the first simple organisms to man. Here, for the first time, the principles of physics are no longer helpful. The stars and planets have yielded the secrets of their history to the physicist; the molecular foundations of living organisms are beginning to be understood; but the *complete* organism—even the simplest and most primitive kind—is incalculably more complicated than any star, or planet, or giant molecule. New insights are needed for the understanding of its structure and evolution. A new law must be found.

The new law was discovered by Charles Darwin more than a century ago. Darwin showed that evolution is the result of a mechanism or "force" in nature, which works on plants and animals slowly, over the course of many generations, to produce changes in their forms. This "force" has no mathematical description; it is not to be found in any textbook of physics, listed alongside the basic forces which control the world of nonliving matter; but, nonetheless, it guides the course of evolution and shapes the forms of living creatures—on this planet and on all planets on which life has arisen—as firmly and as surely as gravity controls the stars and the planets.

Darwin was led to his discovery by observations of plant and animal life carried out between 1832 and 1836, during a voyage around the world on HMS *Beagle*, a British navy vessel assigned to surveying and mapping duties in the southern hemisphere. He had sailed on the *Beagle* as a naturalist, serving without pay and collecting specimens during a journey that carried him around most of the South American continent, to Australia, New Zealand, Africa and many Atlantic and Pacific islands.

Darwin suffered from seasickness throughout the five-year journey, from the day he stepped on board the *Beagle* to the day he quit her decks. He wrote from Brazil, shortly before the end of the voyage, "I loathe . . . the sea and all ships which sail upon it." And when he reached England he never boarded a ship or left his native country again. But throughout the rest of his life Darwin drew on the store of experiences accumulated in that single voyage. Toward the end of his life he wrote, "The voyage of the *Beagle* has been by far the most important event of my life. . . ."

From 1832 to 1835 the ship sailed up and down the coast of South America, and on several occasions during that time Darwin went ashore for long overland journeys of exploration. During these trips ashore Darwin unearthed the evidence which first turned his mind toward evolution. He came upon beds of fossils containing the skeletons of animals that had once roamed the Argentine Pampas but had been extinct for tens of thousands of years.

One of them was a toxodon, "one of the strangest animals ever discovered," the size of a rhinoceros, but with front teeth resembling those of a rodent; another was a "gigantic armadillolike animal" resembling the modern armadillo of South America, but ten times larger.

None of these extinct creatures was the same as any animal now alive on the South American continent; and yet some of them bore a surprising resemblance to existing species. Darwin thought about this "wonderful relationship . . . between the dead and the living"; was it possible that all the animals living on the earth were directly descended from vanished species? Could it be that the passage of vast amounts of time had, in some way, worked the changes between those ancient animals and their modern descendants?

Other facts impressed Darwin later in the voyage. The *Beagle* stopped for a month at the islands of the Galapagos Archipelago, situated on the equator 500 miles west of the coast of South America. During the visit Darwin noticed a "most remarkable feature" of these islands: although they were close to one another, and the climate and soil were the same on all, different kinds of plants and animals lived on them. Some of the islands even contained plants or animals which were to be found on those islands and on no others. On James Island, for example, Darwin found thirty kinds of plants which were exclusively confined to this one island, and were not to be found elsewhere in the Archipelago. He wrote, "I never dreamed that islands, about fifty or sixty miles apart . . . formed of precisely the same rocks, placed under a quite similar climate . . . would have been differently tenanted. . . ."

Darwin puzzled over the "eminently curious" nature of these variations in species; if all forms of life on the earth had been placed here by separate acts of creation, why was the creative force so prodigal in bestowing separate species on each island in the Galapagos?

Another observation led Darwin to the answer. He had also noticed that most of the animals on the islands resembled animals which were peculiar to the neighboring South American continent, and were not to be found in other parts of the world. The significance of this fact did not occur to Darwin until his return from the voyage of the *Beagle* in 1836. Then an explanation occurred to him: a long time ago plants, insects, birds, reptiles and mammals, carried by currents of air and wind, or floating on driftwood, must have reached the Archipelago from the adjacent coast. Isolated from the mainland, they evolved into forms which came to differ more and more, in the course of time, from those of their mainland cousins. Moreover, when the migrant plants and animals first arrived at the Archipelago and became established, they were identical on every island; but gradually, because the islands were isolated and cross-breeding between islands rarely occurred, distinct lines of evolution developed on each. In this way the separate islands acquired their characteristic flora and fauna.

Darwin's reasoning implied that the forms of life could change and

Charles Darwin: A Year Before His Death

evolve with the passage of many generations. By 1838 he was convinced that "such facts as these . . . could only be explained on the supposition that species gradually become modified." He was convinced of the truth of evolution.

These views were contrary to the opinions of many scientists and nearly all laymen, the common view being that every form of life had been specially and independently created. They were contrary also to the views held by Darwin himself when he boarded the *Beagle* in 1832. Yet six years later, Darwin found he could not ignore his own evidence on the Pampas; he had dug fossil skeletons, some closely resembling living forms, out of the ground in Argentina with his own hands; in the Galapagos he had seen with his own eyes the basic resemblance of island animal life to the life on the South American mainland; and he had seen the differences between corresponding species on separate islands in the Archipelago.

But how could Darwin convince a skeptical world that it must accept a theory which violated its basic beliefs? He wrote subsequently in his *Autobiography* regarding his belief in evolution: "The subject haunted me . . . It seemed to me almost useless to endeavor to prove by indirect evidence . . ."

So he sought a direct proof; he looked for a *cause* of evolution—a principle in nature which would make evolution a necessary and inevitable aspect of life, and, at the same time, would explain the different forms which plants and animals had assumed.

Throughout the years following the voyage of the *Beagle*, Darwin's mind turned on the problem of a cause for evolution. Gradually the outlines of a new theory emerged. It is hard to say precisely when Darwin first saw the light; no doubt the truth dawned on him slowly; but by the end of 1838 the new law of nature was clearly formulated in his notebooks. Yet he did not announce it to the world immediately; he knew that his belief in evolution would make him an unpopular figure. "It is like confessing a murder," he wrote later. In order to strengthen his case he first collected every bit of evidence which could bear on the question; he wrote in 1844, "I have read heaps of . . . books, and have never ceased collecting facts"; and in 1858, "I am like Croesus overwhelmed with my riches and facts."

At last, in November 1859, Darwin's theory appeared in print* with the title of

THE ORIGIN OF SPECIES
BY MEANS OF NATURAL SELECTION
OR
THE PRESERVATION OF FAVORED RACES
IN THE STRUGGLE FOR LIFE

Darwin's apprehensions regarding the reception of his theory were confirmed immediately on its publication. The first edition of the *Origin* aroused intense interest; the entire printing sold out on the day of its appearance, and it drew down on the head of its author a storm of vituperation and ridicule such as has never greeted any other work in the history of science. After reading it, his geology professor at Cambridge wrote to him, "I laughed . . . till my sides were almost sore . . . utterly false and grievously mischievous . . . deep in the mire of folly." Other critics were less gentle; an anonymous reviewer wrote in the *Edinburgh Quarterly Review* of Darwin's "rotten fabric of guess and speculation . . . dishonorable to Natural Science."

But the argument set forth in *The Origin of Species* was beautifully simple and clear; its validity should have been apparent to everyone. Darwin began with an almost self-evident set of observations on the nature of life: All living things reproduce themselves; reproduction is

*Darwin might have put off publication and collected facts to the end of his life if an accident of history had not goaded him into action. For years his colleagues had urged him to publish; they had warned him that he would be anticipated if he delayed. In June, 1858, their predictions came true; Darwin received a letter from the naturalist Alfred Russel Wallace proposing a theory of evolution very clearly formulated and identical with Darwin's but arrived at independently. Friends arranged for presentation of a joint paper to the Linnean Society over the names of Darwin and Wallace. Then Darwin set to work in earnest. In 1859, after "13 months and 10 days of hard labor," *The Origin of Species* appeared, and it became clear that Darwin had taken infinitely greater pains than had Wallace to collect evidence for the defense of the theory. It was the mountain of detail in the *Origin,* and Darwin's painstaking analysis of the evidence, that eventually overcame the fierce opposition to the theory and secured its acceptance.

the essence of life; *but the process of reproduction is never perfect.* The offspring in each generation are not exact copies of their parents; brothers and sisters differ from one another; no two individuals in the world are exactly alike, except for identical twins at the moment of birth.

Usually the variations are small; brothers and sisters resemble one another, all human beings look more or less alike, and all elephants look more or less like other elephants.

Yet, Darwin asserted, these small variations are critically important; for, in the struggle for existence, the creature which is distinguished from its brethren by a special trait, giving it an advantage in the competition for food, or in the struggle against the rigors of the climate, or in the fight against the natural enemies of its species—that creature is the one most likely to survive, to reach maturity, *and to reproduce its kind.* Some of the offspring of the favored individual will inherit the advantageous characteristic; a few will possess it to a greater degree than the parent. These individuals are even more likely to survive and to produce offspring.

Thus, through successive generations, the advantageous trait appears with ever-increasing strength in the descendants of the individual who first possessed it.

Not only does the trait become more pronounced in each individual with the passage of successive generations, but the number of individual animals possessing it also increases. For these favored individuals have slightly larger families than the average, because they and their offspring have a greater chance of survival; in each generation they leave behind a greater number of offspring than their less-favored neighbors; their descendants multiply more rapidly than the rest of the population, and in the course of many generations, their progeny replace the progeny of the animals that lack the desirable trait.

In *The Origin of Species,* Darwin gave this process the name by which it is known today: "This principle of preservation or the survival of the fittest, I have called *Natural Selection.*"

Through the action of natural selection, a favorable trait which first appeared as an accidental variation in a single individual will,

This cartoon appeared in 1871 after the publication of Darwin's "Descent of Man."

with the passage of sufficient time, become a pronounced characteristic of the entire species. So the deer become fleet of foot, for the deer which ran fastest in each generation usually escaped their predators and lived to produce the greatest number of progeny for the next generation. So did man become more intelligent, for superior intelligence was of premium value: the intelligent and resourceful hunter was the one most likely to secure food. Thus developed the brain of man; thus, too, in response to other pressures and opportunities in their environments, developed the trunk of the elephant and the neck of the giraffe.

Of course, the incorporation of one new trait does not create an entirely new animal. But if we count all the births which occur to a single species over the face of the globe in one year, an enormous number of variations will appear in this multitude of young creatures. On all these variations the same process of selection works steadily, preserving for future generations the new traits which give strength to the species, and eliminating those which lend weakness. The changes may be imperceptible from one generation to the next, but over the course of many generations the accumulation of many favorable variations, each slight in itself, completely transforms the animal. According to Darwin,

> Natural selection is daily and hourly scrutinising, throughout the world, the slightest variations, rejecting those that are bad, preserving and adding up all that are good; silently and insensibly working at the improvement of each organic being in relation to its . . . conditions of life.

Natural selection molds the forms of life. Under its continuing action the shapes of animals change with time; old species disappear in response to changing conditions, and new ones arise. Few of the species of animals which roamed the face of the earth 10 million years ago still exist today, and few of those existing today will survive 10 million years hence. To quote again from *The Origin of Species*; ". . . Not one living species will transmit its unaltered likeness to a distant futurity." But natural selection works its effects subtly. Its influence is not felt in one individual or in his immediate descendants. A thousand generations may elapse before a change becomes noticeable; in man that amounts to 20,000 years. Yet, ever since Rutherford measured the age of the earth, we have known that enough time is available. Our planet has existed for billions of years; that is the secret strength of Darwin's theory. "We have almost unlimited time," he wrote in 1858, in explaining how the slightest variations in the form of an animal can grow, through their effect on the probability of producing progeny, until, after the passage of "millions on millions of generations," great changes are effected. And in the *Origin:*

> The mind cannot grasp the full meaning of the term of even a million years; it cannot add up and perceive the full effects of many slight variations, accumulated during an almost infinite number of generations. . . . We see nothing of these slow changes in progress, until the hand of time has marked the lapse of ages, and then . . . we see only that the forms of life are now different from what they formerly were.

Darwin's critics were not accustomed to thinking in terms of millions of generations and tens of millions of years; they accused him of proposing that natural selection could convert "an oyster into an orangutan" or "tadpoles into philosophers"; they taunted him with his inability to supply the missing link—the animal caught midway in the transition from one species to another. A British magazine wrote in 1861, "We defy any one, from Mr. Darwin downwards, to show us the link between the fish and the man. Let them catch a mermaid. . . ."

The hapless naturalist could not oblige them. Throughout the long battle for the acceptance of his views, Darwin was plagued continually by his inability to compress the time scale of nature and demonstrate a transformation of species to his critics. Had he known it, an example was at hand which would have provided him with the proof he needed. The case was an exceedingly rare one, in which a major evolutionary change occurred in the brief interval of fifty years.

The animal which underwent the transformation was a member of the insect world, the humble Peppered Moth, found in abundance throughout England. In the nineteenth century two varieties of this moth were known. One possessed a speckled coloration, blending perfectly into the background of lichen-covered tree trunks, which provided its normal resting place. The other variety was dark, almost black in color, and stood out conspicuously against the light background of lichens and bark. The speckled variety, known as the Peppered Moth, was the commoner form; the dark variety was readily picked off and eaten by birds, and was relatively rare.

During the course of the nineteenth century, soot progressively darkened the tree trunks of the English Midlands. As much as two tons of soot fell each day on every square mile of some industrial towns in that area. The speckled coloration of the Peppered Moth, which must originally have appeared as a chance mutation, had been developed and refined by the action of natural selection over many generations, because of its favorable effects in the struggle for survival. This same characteristic, because of a change in the environment, in this case wrought by man, now placed its possessor at a disadvantage; the Peppered Moth stood out clearly against the background of the soot-covered tree trunks, and was detected with ease by the birds of the region. The black moth, on the other hand, blended well into the new background; the trait which had formerly been unfavorable was now favorable, enhancing the chances of survival to maturity and the production of offspring. The black moth, once a rare variety, multiplied in number until it became the dominant form. The change was dramatic and swift; the first recorded capture of a dark moth took place in Manchester in 1848, and by 1900 the dark moth outnumbered the speckled variety by 99 to 1.

Even if Darwin had been able to produce the example of the Peppered Moth, it is doubtful if he would have stilled the voices of criticism, for the objections to his theory of evolution were not raised on rational grounds alone. There was also an emotional reaction to the implications of the theory for the descent of man. In the *Origin*, Darwin had deliberately avoided discussion of man's ancestry; while the book was in preparation he had written to a friend, "You ask whether I shall discuss man. I think I shall avoid the whole question. . . ." Darwin's critics were quick to supply the missing discussion: Darwin asserted that the forces of nature, acting through the struggle for survival, work continuously for the improvement of all forms of life; it followed that each animal now on the face of the earth must be descended from a related but more primitive ancestor. What animals provided a clue to man's primitive ancestry? Monkeys and apes were less advanced than man, yet closer to him in form and intelligence than any other creature; they represented primitive forms of the human being. The ridiculous monkey and the brutelike gorilla resembled man's ancestors.

Many articulate defenders of man's noble heritage entered the lists against the rash and blasphemous scientist. One of the most prominent and eloquent anti-Darwinians was Samuel Wilberforce, Bishop of Oxford. On June 29, 1860, six months after the publication of the *Origin*, 700 people crowded into a hall in Oxford University to hear Bishop Wilberforce debate the merits of Darwin's theory with the biologist, Thomas Huxley, who had become Darwin's most ardent supporter. Toward the end of the debate, Bishop Wilberforce turned to Huxley and asked, "Was it through his grandmother or his grandfather that he claimed his descent from a monkey?"

Huxley's reply is one of the most famous ripostes in the history of science. Whispering to his neighbor, "The Lord hath delivered him unto my hands," he rose and said, "If I am asked whether I would choose to be descended from the poor animal, of low intelligence and stooping gait, who grins and chatters as we pass—or from a man, endowed with great ability and a splendid position, who would use these gifts to discredit and crush humble seekers after truth, I hesitate what answer to make."

The arguments over Darwin's views gradually subsided through the decade of the 1860s. They erupted again when the *Descent of Man* was published in 1871. In this book, Darwin presented his views on man's origin and history, confirming the darkest suspicions of his critics by setting forth evidence linking man and the apes to a common ancestor. But by Darwin's death in 1882, his theories were widely accepted in the scientific world, and had made a substantial impact on the thinking of all men. Today the basic concepts of the Darwinian theory of evolution have few opponents.

AN EXAMPLE OF NATURAL SELECTION. Usually the pace of evolutionary changes is too slow to be observed within the span of one lifetime, but in the case of the Peppered Moth of England a major evolutionary change occurred in fifty years. Originally this moth possessed speckled gray-and-white markings, which blended perfectly into the lichen-covered trunks of trees. A speckled specimen appears near the center of the photograph *(above)*, barely visible against the tree trunk. In the same photograph appears a second variety, very dark in color because its body chemistry manufactured an excessive amount of melanin. In older times the dark specimens were rare because they stood out easily from the barks of trees and were readily detected and eaten by birds.

In the aftermath of the Industrial Revolution tree trunks throughout large areas of England were blackened by soot, against which the speckled moth stood out conspicuously *(above)*. The dark variety, which appears at the bottom of this photograph, is nearly invisible against the soot-blackened tree. This change in environment led to the rapid decimation of the Peppered Moth; the dark variety rapidly became the dominant species, outnumbering the original form of this moth by 100 to 1 in a census taken in 1900.

Industrial changes in the environment converted a formerly unfavorable mutation, the excess of melanin, to a favorable mutation in the course of only fifty years.

CHANCE AND DESIGN. Darwin's theory of evolution by natural selection provides a generally satisfactory explanation of the forms of life present on the earth, and is accepted by most scientists. However, it is difficult to see how this process, based on a succession of random mutations, can account for such highly specialized forms as the tree hopper *opposite*. This thorn-shaped insect clings to the stem of a rosebush, escaping the eye of birds. The survival value of its peculiar shape is clear, but how that shape evolved by chance mutations is less clear.

The photographs on pages 242-243 *(overleaf)* show another remarkable similarity in nature, this time between an owl's eyes and two spots on the wings of the caligo butterfly. If Darwin's theory is correct, the resemblance must be the result of a series of imperceptible changes over many generations, which started when a butterfly acquired two spots of pigmentation on its wings as a result of a chance mutation. The dark spots looked enough like the eyes of an owl to be frightening to some birds. In each generation, according to the theory of natural selection, the butterflies with darker and rounder spots than their parents were more likely to escape the attacks of their predators; thus their special trait of an eye-shaped pigmentation was more likely to be preserved and passed on to the next generation. Over the course of many generations, the crude resemblance became refined into a nearly perfect imitation of an owl's eyes.

Prior to the publication of Darwin's "Origin of Species," special creation was the only explanation available for such intricately designed forms of life as the tree hopper and the caligo butterfly.

16 DNA AND DARWIN

Throughout the years in which Darwin's ideas were winning increasing acceptance, one point remained obscure. What was the origin of the variations from one individual to another, which provided the raw material for natural selection? Regarding these variations, which played so essential a role in his theory, Darwin could only say helplessly, "We are profoundly ignorant of the cause of each slight variation or individual difference. . . .[They] seem to us in our ignorance to arise spontaneously." The ignorance was not fully dispelled until 1953, nearly a century after the publication of the *Origin*, when it became clear how the basic characteristics of the individual are passed from generation to generation. These characteristics reside in the molecule called DNA, which is found in the cells of every living organism on the earth. The DNA molecule is a long chain of the smaller molecules called nucleotides, which are arranged in a sequence special to each organism. As we have noted, no two individuals in the world, except identical twins, have the same sequence of nucleotides in their DNA. This sequence determines which proteins will be assembled in the cells of the body; and the proteins, in turn, control the body chemistry and all the traits of the individual. Thus, the DNA in every creature contains the master plan for that creature.

We now know that occasionally some of the nucleotides in the DNA molecule are damaged, altered or removed entirely from the molecule, so that the master plan is changed. The damage or alteration may affect only one nucleotide in the long chain—a chain which may, in human cells, stretch out over a billion nucleotides. Nevertheless, the change in a single nucleotide may be critically important, for the proteins in the cell are assembled out of amino acids in a sequence that follows the order of the nucleotides in the DNA. Damage to one of these nucleotides, or replacement of one nucleotide by another of a different type, will lead to the assembly of a different protein, in which, at one point along the chain of amino acids making up the protein, the wrong kind of amino acid is located.

Sometimes the modified protein is able to play its normal role in the chemistry of the cell. At other times, when the improperly placed amino acid is located at a critical site, the effectiveness of the protein is destroyed.

When a modified DNA and the modified protein produced by it are situated in an ordinary cell of the body, the abnormal cell is soon replaced by the growth of new cells and the effect of the change in the DNA molecule quickly disappears. However, in one type of cell in the body a change in the sequence of nucleotides in the DNA molecule may have permanent and serious consequences. This is the germ cell — sperm in the male and ovum in the female. Like all other cells, the germ cell contains its set of DNA molecules with the master plan for the development of the individual. When the sperm and ovum unite to form a fertilized egg, every organ in the body of the mature individual subsequently develops out of the egg by repeated cell division, following a new, joint master plan provided by the combination of the DNA molecules from the sperm and the ovum participating in the union. If the DNA in one of these two germ cells has been damaged or altered, the effect of the change will appear in every cell of the body of the new individual. Moreover, it will be transmitted to the offspring of this individual in the next generation, and in every generation thereafter. All the descendants of that individual, down through the corridors of time, will bear the mark of the change in the sequence of nucleotides in the ancestral DNA.

A modification of the nucleotides in the DNA of a germ cell is called a *mutation*. Mutations are changes in the body chemistry of the individual *which are transmitted to his progeny*. They are the inheritable variations which form the basis of Darwin's theory of evolution.

Some mutations change the chemistry of the body in a way which improves the chances of the individual for survival; these are called

favorable mutations. The individuals possessing a favorable mutation are always the ones most likely to propagate their species; from
generation to generation their number steadily increases, and in the
course of many generations the favorable mutations spread
throughout the population. Mutations may also be *unfavorable*,
diminishing the chance of survival to maturity and, therefore, the
chance of producing offspring. These mutations are gradually weed-
ed out of the population. That is the way in which evolution works:
By the pruning away of unfavorable mutations simultaneously with
the strengthening of favorable ones.

What is the cause of mutations? What can damage or modify the
sequence of nucleotides in the germ cells of an organism? When we
know the answer to this question, we are close to an understanding
of the cause of evolution.

One cause of mutations lies in the organism itself. Occasionally an
error—an imperfection—occurs in the copying process by which the
DNA molecule reproduces itself. At the start of the process of
reproduction, just prior to the division of a cell into two daughter
cells, the double-stranded DNA in the parent cell unwinds and
separates into two single strands. Each strand gathers new
nucleotides from the pool of nucleotides floating in the cell to form a
new, double-stranded DNA duplicating the original. It is at this point
that the copy error may occur. One of the newly added nucleotides
may be of the wrong kind; that is, it fails to match its counterpart in
the existing strand. As a result, when the assembly of the two DNA
molecules has been completed, the sequence of nucleotides in one of
the daughter DNAs differs from the sequence of nucleotides in the
parent DNA. That daughter DNA has suffered a mutation.

Copy errors are not the only source of mutations. The DNA
molecule can also be altered by chemicals, if they enter the blood
stream. Mustard gas, the poison gas of the First World War, is effective in this way. LSD appears to cause serious damage to the DNA
molecule. And DNA can be altered by particles or radiation that are
energetic enough to penetrate the body. The physician's x-ray
machine is one source of penetrating radiation. Nuclear bomb explo-

sions are another. In addition to these man-made sources, there are also cosmic rays—particles produced by unknown forces in distant regions of the universe—which bombard the earth from all directions. Occasionally, either a cosmic ray or radiation produced by a man-made source will pass through a germ cell, even if the cell is buried deeply in the body, and will disrupt the normal sequence of nucleotides in its DNA, producing a mutation.

Which of these sources is the prime cause of mutations? Mustard gas, x-ray machines and nuclear bombs are recent products of man's ingenuity; they have not been with us long enough to have had an appreciable influence on the course of evolution. But copy errors have occurred since the beginning of life on earth, and cosmic rays probably have existed since the beginning of time. We can guess that these latter two sources have played roles of comparable importance in the past history of the evolutionary process. Whether the rate of mutation and the future rate of evolutionary change will be increased appreciably by the other, man-made sources of mutation, and whether the augmentation will improve or weaken the human species — these are open questions.

17 THE ASCENT OF MAN

In the course of 3 billion years, life on the earth evolved from a soup of organic molecules to the carnival of animals that now plays across the face of the planet. Among these animals is man. By what sequence of events did he arise out of a broth of DNA and proteins? What circumstances guided the course of evolution from the first primitive organisms to the highest expression of life in the form of the human being?

The history of these events probably began with the appearance of the first self-copying molecules in the waters of the earth. These molecules were similar to DNA, and may have been identical with it. The waters also contained amino acids. In some way, which we have not yet been able to reconstruct in the laboratory, those self-copying molecules developed the ability to serve as guides for the assembly of amino acids into proteins. At first the DNA molecules were short strands containing only a few nucleotides, and could assemble only simple proteins. In the course of time they evolved into longer chains, capable of assembling complex proteins of many kinds. Some of these proteins were primitive enzymes; they hastened the chemical reactions which led to the growth and reproduction of DNA. Other proteins were structural; they were of a kind out of which cell walls were formed.

With the appearance of the structural proteins, a new advance became possible in the organization of living matter. The DNA molecule came to reside in the center of a cell, whose wall was a porous membrane which permitted small molecules, such as amino acids and nucleotides, to pass through from the surrounding fluid to the interior, but did not permit the larger molecules, such as DNA and the proteins assembled under its control, to pass out again in the reverse direction.

The primitive development of the cell, concentrating in the vicinity of the DNA all the chemicals needed for growth and reproduction, marked the greatest single step ever taken in the course of evolution. Several hundred million or, possibly, a billion years must have been required for the evolution of the cell; but once it appeared, this efficient form of life must have spread rapidly throughout the waters of the earth, engulfing and replacing all the cell-less molecules which preceded it.

We can assume that in a relatively short time—perhaps within 100 million years—the one-celled organism evolved into a colony of cells. With the further passage of time, groups of cells within those colonies assumed specialized functions of food-gathering, digestion, the structural features of an outer skin, and so on; thus began the stage of evolution leading to the complex, many-celled creatures which dominate life today.

The fossil record contains no trace of these preliminary stages in the development of many-celled organisms. The first clues to the existence of relatively advanced forms of life consist of a few barely discernible tracks, presumably made in the primeval slime by soft, wriggling, wormlike animals. These are found in rocks about one billion years old. Somewhat later, well-defined worm burrows appear in the record. These meager remains are the earliest traces of many-celled animal life on the planet.

Little else appears in the fossil record during those first several billion years. One of the mysteries in the study of life is the fact that suddenly, in rocks 600 million years old, the record explodes in a profusion of living forms. A great variety of animals appears in the record at that time. Perhaps the forms of life were nearly as numerous and populous just prior to this magical date, but left no trace of their existence because they lacked the body armor which is most easily preserved.

Somewhat more than 400 million years ago an event occurred which is of great consequence for the development of man. There appeared, for the first time, a new kind of creature—one with an internal skeleton and a backbone. This animal—the vertebrate—evolved out of a wormlike ancestor resembling the modern lancelet, a small, translucent creature, lacking fins and jaws, but possessing gills and, most important, a primitive version of the backbone.

Among the descendants of the first vertebrates were the fishes. Some of the early fishes contained crude lungs for gulping air at the surface of the water, as well as gills. These lungs were lost or converted to other uses in most instances, but in some forms of fish, perhaps those living in small bodies of water such as ponds and tidepools, the lungs came into frequent use. Whenever a drought developed and the water level in the ponds dropped, the fish with the best lung capacity survived where others perished. They lived to produce progeny which inherited their superior capacity for breathing air. In this way, an efficient lung evolved gradually among the fish inhabiting shallow bodies of water.

Some of the air-breathing fish were doubly favored in possessing strong fins which enabled them to waddle over the land from one pond to another in search of water. By a slow accumulation of favorable mutations, the muscle and bone of the fin gradually changed into a form suitable for walking on land. In this way, the fin evolved into a leg. The metamorphosis took place over a period of perhaps 50 million years, and a like number of generations. The result was a four-legged, air-breathing animal, known as the amphibian.

The amphibian was still tied to the water, because its skin required frequent moistening; also, its eggs, like those of fish, lacked a hard casing. If deposited on land they dried out and the embryo died. Therefore the eggs of the amphibian had to be laid in water or in moist places.

The amphibians were born in water, lived most of their adult lives near water, and almost always returned to the water to lay their eggs. For fifty million years they flourished on shores and river banks. Some became large, aggressive carnivores as much as 10 feet in length, fearing no other animals of their time. The amphibians attained the peak of their size 250 million years ago, and thereafter they declined.

Today their common descendants are the diminutive frog, toad and salamander.

In the course of time, some of the ancient amphibians, again by the chance occurrence of a succession of favorable mutations, developed the ability to lay their eggs on land. These eggs were encased in a firm, leathery shell, which retained moisture and provided the embryo with its own private pool of fluid. Other mutations led to a tough, leathery hide, which preserved the water in their bodies without the need for continual immersion. Such creatures were completely emancipated from the water. They were the first reptiles.

The reptiles marked a very successful step in evolution, for they had access to rich resources of food previously denied to the fishes and the amphibians. The reptiles flourished and developed into a great variety of forms, including the ancestors of every land animal with a backbone now on this earth. They reached their evolutionary zenith in the dinosaurs. These animals ruled the earth for 100 million years. They displayed an extraordinary vigor, evolving into such extreme forms as the giant vegetarian swamp-dweller, *Brontosaurus*, 70 feet long and weighing 30 tons; and the meat-eating *Tyrannosaurus rex*, 40 feet high, with a 4-foot skull filled with daggerlike teeth—unquestionably the fiercest land-living predator the world had ever seen.

Two hundred million years ago, somewhat before the appearance of the first dinosaurs, another branch of the reptile class veered off on an entirely different course. This particular group may have lived in places on the edge of the temperate zone where the weather was relatively severe. Through the action of natural selection on chance variations, the new branch of the reptiles acquired a set of traits which fitted them uniquely for survival in a rigorous climate. They developed the rudimentary characteristics of a warm-blooded animal. The naked scaly skin of the reptile was replaced in these animals by insulating coats of hair and fur which kept them warm in low temperatures, while sweat glands under the skin, controlled by an internal thermostat, cooled the body by evaporation when the temperature rose too high.

These traits developed slowly over the course of tens of millions of years. Other changes took place at the same time. Many four-footed reptiles had a clumsy, sprawling posture with the legs spread out from the trunk; rapid movement was impossible with a skeleton so constructed. In the new line of evolution the legs pulled in beneath the body, raising it from the ground and permitting a fast, running gait. Important modifications appeared in the teeth: near the front of the mouth were two large canines suitable for tearing off pieces of the prey; behind these were cusped teeth resembling molars, for cutting and grinding the food down to a smaller size. A quick replenishment of energy and a high level of activity are possible with such teeth, in contrast to the postprandial stupor of the reptile that has swallowed its prey whole.

These animals, resembling a cross between a lizard and a dog, constituted the ancestral stock of the mammals.

Modern mammals possess other characteristics, in addition to warm-bloodedness, which distinguish them from their reptilian forebears; the most important among these is an exceptionally effective means, unique to mammals, of caring for their young. Reptiles lay their eggs and commonly display no further interest in the fortune of their progeny. Birds are better parents; they care for their eggs, but the defenseless, unhatched embryo is, nonetheless, often the victim of egg-hungry creatures; and, moreover, the newly hatched fledgling must be left to the mercies of predators while its parents search for food. The mammalian mother nourishes the developing embryos of the unborn young inside her body, where they are well protected against hostile elements in the environment; after birth, she nourishes her young with her milk, secreted by the glands which have given the mammals their name; and she continues to care for the young a long time thereafter, until they are able to fend for themselves. The mammals make more effective provisions for the survival of their young than any other animal, thereby securing a very great advantage in the competition for the propagation of their species.

In spite of these special talents, the mammals remained subordinate to the dinosaurs for more than 100 million years—small, furry animals, inconspicuous, keeping out of sight of the rapacious reptiles by living in the trees or in the grasses.

But seventy million years ago, the dinosaurs died out. The reasons for their disappearance are still obscure. It is likely that their downfall was the consequence of a worldwide change in climate, which they were ill-equipped to survive. Dinosaurs, like all reptiles, were cold-blooded animals; that is, they lacked the internal heat controls which could maintain the temperature of the body at a constant level regardless of the rigors of the climate. We know that the period during which they disappeared was marked by repeated upheavals of the earth's crust, in which many new mountain ranges were formed. The Rocky Mountains were among the ranges created in these upheavals. Most probably, the upward thrust of huge masses of rock disrupted the flow of currents of air around the globe; perhaps the climate of the temperate zone was changed in this manner from one of uniform warmth and humidity, agreeable to a cold-blooded animal, to a climate marked by major changes of temperature from season to season.

As the population of dinosaurs dwindled, the mammals came down from the trees and up from their burrows in the ground, and they inherited the earth. Quickly they spread out across all the continents. Within 20 million years, the basic mammalian stock evolved into the forebears of most of the mammals with which we are familiar today—bats, elephants, horses, whales and many others.

But one group of mammals remained in the trees. These mammals—the primates—were singled out, by the circumstance of their tree-dwelling existence, to be the ancestors of man. They were small, insect-eating animals, the size of a squirrel, and similar in appearance to the modern tree shrew of Borneo. Man owes his remarkable brain to the fact that these animals required two physical attributes for survival in their arboreal habitat: First, they needed hands and an opposable thumb for securing a tight hold on branches; and second, they needed sharp binocular vision to judge the distances to nearby branches. In the competition for survival among primitive tree-dwelling mammals, 100 million years ago, those who possessed these characteristics in the highest degree were favored. They were the individuals most likely to survive and to produce offspring. Through successive generations the desirable traits of a well-developed hand and keen vision, passed on from parents to offspring, were steadily refined and strengthened. By 50 million B.C. they already appeared in advanced form in the animals from which the modern tree shrew, lemur and tarsier are descended. They became even better developed in some of the immediate descendants of these animals, under the continued pressure of the struggle for survival in the trees. Gradually, the evolutionary trends established by the requirements of life in the trees transformed some of these early primates into animals resembling the monkey.

Animals with hands also had the potential capacity to exercise rudimentary manual skills; when this potential was combined with the development of the associated brain centers, such animals had, almost by accident, the ability to use tools. For those who had this ability, great value became attached to the mental capacity for the remembrance of the usage of tools in the past, and for the planning of their use in the future; thus, by the action of natural selection on a succession of chance mutations, those centers of the brain developed and expanded in which past experiences were stored and future actions were contemplated. These mental qualities proved to be of great value in meeting the general problems of survival. As a result, the brain evolved and expanded under the continued pressure of the struggle for existence. It doubled in size in 10 million years, and nearly doubled again in the next million. Thus was the line of ascent leading to man firmly established.

By this chain of evidence and theory, the distinguishing characteristic of the human condition — intelligence — may be traced back to the accidental circumstance of a tree-dwelling ancestry.

The path of evolution stretches further back into time — from the tree-dwelling forebears of man to the first mammal; then to a doglike reptile of a kind that no longer exists; to the first vertebrate; from the vertebrates to a succession of soft-bodied animals lost in the sands of history; then across the threshold of life into the world of nonliving matter; and finally, many billions of years ago, long before the solar system existed, into the parent cloud of hydrogen.

There is grandeur in this view of life, with its several powers, having been originally breathed by the Creator into a few forms or into one; and that, whilst this planet has gone cycling on according to the fixed laws of gravity, from so simple a beginning endless forms most beautiful and most wonderful have been and are being evolved.

—CHARLES DARWIN
The Origin of Species

THE HISTORY
OF LIFE

EARLY FORMS OF LIFE. Among the oldest residues of living organisms are the stromatolites, deposits left by primitive algae approximately 3 billion years ago. The photograph above shows an algal stromatolite found in the Precambrian rocks of the Medicine Bow Mountains in Wyoming. On the opposite page are other remains of early life. The residues and simple fossils on these pages are typical of the earliest traces of life discovered thus far.

The fossilized remains of rod-shaped bacteria 3 billion years old *(left and below).*

Burrows made by worms that lived approximately 1 billion years ago *(below).*

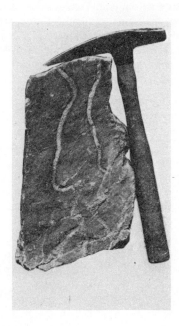

THE EVOLUTION OF THE VERTEBRATES. Four hundred million years ago the first animal with a backbone — a primitive fish — appeared on the earth. Some of the early fishes developed the ability to breathe air and to crawl up onto the land. From these creatures developed the amphibians and then the reptiles. Reptiles, the first backboned animals to be completely emancipated from the sea, spread out over the land and flourished there for 150 million years. Some of the reptiles were the ancestors of the modern snake, lizard and turtle; others evolved into the dinosaurs. One line led to the ancestors of the modern birds and, about 200 million years ago, other branches evolved into the mammals. Both birds and mammals were distinguished from the reptiles in one critical way: they were warm-blooded. This characteristic enabled them to withstand extremes of hot and cold in the external environment. Sixty million years ago, when the dinosaurs suddenly disappeared, probably because of severe changes in climate, the mammals inherited the earth and evolved into myriad forms. One branch of the mammals — the primates — were the ancestors of man.

Bir

Early D

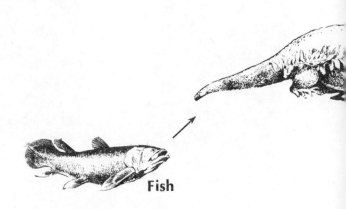

Fish

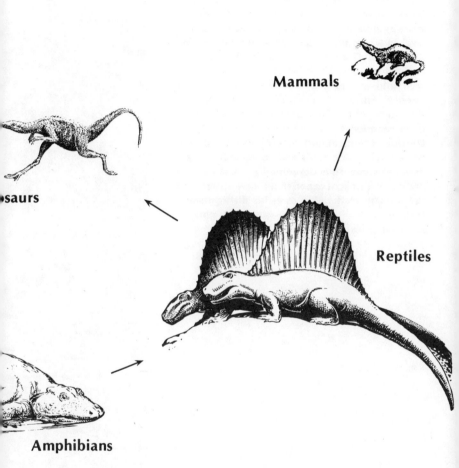

Mammals

saurs

Reptiles

Amphibians

A RELATIVE OF MAN. The tarsier *(opposite)* resembles the small tree-dwelling mammals that were the progenitors of man. In order to survive in the trees these animals needed grasping hands and binocular vision. The possession of these characteristics stimulated the development of the brain centers that coordinated sight and touch. In the struggle for existence 50 million years ago, those creatures were selected for survival in whom these traits were the most highly developed. Gradually the tarsier-like animals evolved into the ancestor of the monkey, and then into an ape-like creature which was the forebear of man. Manual dexterity and the development of the associated brain centers, traits required for an arboreal existence, gave their possessors, almost by accident, the potential ability to use tools and to remember past usage of tools. Thus developed the parts of the brain in which past experiences were stored and future plans made. Man owes his distinguishing feature — his intelligence — to the circumstance that his ancestors lived in trees.

ADDITIONAL READING

A fuller account of the birth and death of stars and methods of astronomical measurement may be found in *Astronomy: Fundamentals and Frontiers* by Robert Jastrow and Malcolm H. Thompson. The evidence for the expanding universe and the cosmic explosion is explained in more detail, but without mathematics, in *God and the Astronomers* by Robert Jastrow. *The First Three Minutes* by Steven Weinberg contains a more technical but eminently readable and informative account of the same matters. *In the Center of Immensities*, by the distinguished British astronomer Bernard Lovell, covers the new scientific cosmology with greater sensitivity to its philosophical and theological implications than is to be found in most contemporary writing on the subject.

An excellent account of the play of events and personalities leading to the discovery of the double-helix structure of DNA is to be found in Horace Freeland Judson's *The Eighth Day of Creation*, together with an extremely detailed but relatively non-technical discussion of the mechanism by which the DNA molecule controls the manufacture of proteins.

The Meaning of Evolution by George Gaylord Simpson gives a clear explanation of Darwin's theory of evolution, illustrated by many examples from the fossil record. Darwin's own style is lucid and delightful, and can be sampled with great profit by reading *The Origin of the Species*, in particular Chapter 4 on "Natural Selection; or the Survival of the Fittest" and the concluding Chapter 15. The quotation on p. 255 of this book is taken from Chapter 15. Darwin's encounter with Lord Kelvin, and other interesting information about Darwin's life and work, can be found in *Darwin's Century* by Loren

Eiseley. The richest sources of information regarding Darwin's personality, character and thinking are to be found in his *Autobiography*, and in *The Life and Letters of Charles Darwin* and *More Letters of Charles Darwin*, edited by his son Francis Darwin. Edwin H. Colbert's *Evolution of the Vertebrates* is a professional but well written and readable account of the recent history of life, from the first fishes to man.

PICTURE CREDITS

INDEX

About the Author

Robert Jastrow was born in New York City. He received his B.A., M.A. and Ph.D. degrees from Columbia University in theoretical physics. He was a postdoctoral fellow at Leiden University, the University of California at Berkeley, and the Institute for Advanced Study in Princeton, taught physics at Yale, and joined the National Aeronautics and Space Administration at the time of its formation.

His initial research experience was in the field of nuclear physics, especially the properties of the basic nuclear force. His later work has been in planetary science, atmospheric physics, and weather and climate prediction.

Dr. Jastrow is the founder and director of the Goddard Institute for Space Studies, located in New York, which conducts research in astrophysics, cosmology and planetary science under NASA auspices. Dr. Jastrow is also Professor of Astronomy and Geology at Columbia University and Professor of Earth Science at Dartmouth College.

Dr. Jastrow has received the Columbia University Medal for Excellence, the Arthur Flemming Award for outstanding service in the U.S. Government and the NASA Medal for Exceptional Scientific Achievement. He is the author of *Until the Sun Dies* and *God and the Astronomers*, co-author of *Astronomy: Fundamentals and Frontiers* with Malcolm Thompson, and editor of *The Exploration of Space*, *The Origin of the Solar System* and *The Venus Atmosphere*.